Biofuel Cropping Systems

Choosing appropriate practices and policies for biofuel production requires an understanding of how soils, climate, farm types, infrastructure, markets, and social organisation affect the establishment and performance of these crops.

This book highlights land-use dynamics, cultivation practices related to conversion, and wider impacts. It explores how biofuel production chain development is steered by emerging technologies and management practices and how both can be influenced by effective policies designed to encourage sustainable biofuel production.

Also, the book highlights major biofuel production chains including:

- Cane cultivation in Brazil
- Corn ethanol in the USA
- Wheat and rapeseed in Europe
- Oil palm in the Far East
- Cane in Asia and Africa
- SRC and other lignocellulosic crops

In each case, the development, cropping systems, and impacts are discussed; additionally, system dynamics are shown and lessons are drawn for the way things could or should change.

Biofuel Cropping Systems is a vital resource for all those who want to understand the way biofuels are produced and how they impact other elements of land use and especially how improvements can be made. It is a handbook for students, biofuel producers, researchers, and policymakers in energy and agriculture.

J.W.A. Langeveld holds an MS in tropical agronomy. After a career of 20 years of research in public institutions, he founded a private company for research on sustainable biomass (Biomass Research) in 2008. He is an expert on land use and cropping systems analysis. His major focus is on sustainable biomass production and its utilisation in biofuels and biobased materials.

John Dixon is Principal Adviser – Research Programmes, Australian Centre for International Agricultural Research. He has over 30 years of developing country experience with agricultural research and development, including cropping systems, economics and natural resource management with the CGIAR system and the FAO UN.

Herman van Keulen was trained as a soil scientist and production ecologist at Wageningen University. During his career, Herman was recognized internationally as a giant in his field. His knowledge of crop production around the world was encyclopaedic and his readiness to share his wisdom was a constant support for many around him. Sadly, Herman passed away before the publication of the book. He will be missed dearly by his wife and two sons, and also by his former colleagues and numerous ex-students.

Biofuel Cropping Systems
Carbon, land and food

Edited by J.W.A. Langeveld, John Dixon
and Herman van Keulen

earthscan
from Routledge

First edition published 2014 by Routledge

2 Park Square, Milton Park, Abingdon, Oxforshire OX14 4RN
711 Third Avenue, New York, NY 10017

*Routledge is an imprint of the Taylor & Francis Group,
an informa business*

First issued in paperback 2018

British Library Cataloguing-in-Publication Data
A catalogue record for this book is available from
the British Library

Library of Congress Cataloging-in-Publication Data
A catalog record has been requested for this book

ISBN13: 978-0-415-53953-1 (hbk)
ISBN13: 978-1-138-36417-2 (pbk)

Typeset in Sabon
by Apex CoVantage, LLC

Dedication

This book is dedicated to the hundreds of farmers from Africa, Asia, Australia, Europe and America we met during our careers.

And to:

Johannes J. Langeveld, Abcoude, the Netherlands
Keith Dixon, Wandoan, Australia

In appreciation of the dedicated, intelligent and innovative way they engage(d) in farming, and in making sustainable use of valuable resources.

Contents

List of Figures and Tables

Figures

Tables

List of acronyms

ADB	Asian Development Bank
AGLINK/ COSIMO	Worldwide Agribusiness Linkage Program + COmmodity SImulation MOdel
a.i.	Active Ingredient
B3	Fuel mixture with 3% biodiesel by volume
B7.5	Fuel mixture with 7.5% biodiesel by volume
B10	Fuel mixture with 10% biodiesel by volume
BAU	Business As Usual
BP	Biomass Productivity
BRICS	Brazil, Russian Federation, India, China, South Africa
C3	Metabolic pathway for carbon fixation producing a molecule with 3 C-atoms
C4	Metabolic pathway for carbon fixation producing a molecule with 4 C-atoms
CAP	Common Agricultural Policy
CAPRI	Common Agricultural Policy Regionalised Impact
CARB	California Air Resources Board
CARD	Center for Agricultural and Rural Development
CFS	Committee on World Food Security
CGE	Computable General Equilibrium
CHP	Combined Heat and Power
COD	Chemical Oxygen Demand
COSIMO	COmmodity SImulation Model
CPO	Crude Palm Oil
DDGS	Dried Distillers Grains with Solubles
E10	Fuel mixture with 10% ethanol by volume
E85	Fuel mixture with 85% ethanol by volume
E-CROP	Energy Crop Simulation Model
EFB	Empty Fruit Bunches
EIA	Energy Information Administration
EPA	Environmental Protection Agency
ESIM	European Simulation Model
ETBE	Ethyl *Tert*-Butyl Ether
FAO	Food and Agricultural Organization of the United Nations
FAPRI	Food and Agriculture Policy Research Institute
FAPRI-ISU	Food and Agricultural Policy Research Institute, Iowa State University
FFB	Fresh Fruit Bunches
GDP	Gross Domestic Product
GEMIS	Global Emission Model for Integrated Systems
GHG	Greenhouse Gas
GREET	Greenhouse Gases, Regulated Emissions, Energy use in Transportation

GTAP	Global Trade Analysis Project
GWP	Global warming potential
HY	High Yields scenario
ICRISAT	International Crops Research Institute for the Semi-Arid Tropics
IEA	International Energy Agency
IFPRI	International Food Policy Research Institute
IIASA	International Institute for Applied Systems Analysis
ILUC	Indirect Land Use Change
IMPACT	International Model for Policy Analysis of Agricultural Commodities and Trade
INMAS	Intensifikasi Massal
IPCC	International Panel on Climate Change
JRC	Joint Research Centre of the European Commission
LCA	Life Cycle Assessment
LEI	Agricultural Economics Research Institute
MAR	Mean Annual Rainfall
MCI	Multiple Cropping Index
MTBE	Methyl *Tert*-Butyl Ether
MVO	Product Board for Margarine, Fats and Oils
N_2O	Nitrous Oxide
OECD	Organisation for Economic Co-operation and Development
PE	Partial Equilibrium
PEP	PhosphoEnolPyruvate
POME	Palm Oil Mill Effluent
RAI	Responsible Agro Investments
RBD	Refining, Bleaching and Deodorizing
RED	Renewable Energy Directive
RFA	Renewable Fuels Agency
RFS	Renewable Fuel Standard
RTFO	Renewable Transport Fuel Obligation
SADC	Southern African Development Community
SOM	Soil Organic Matter
SRC	Short Rotation Coppice
SRF	Short Rotation Forest
TS	Total Solids
UN	United Nations
VFA	Volatile Fatty Acids
VS	Volatile Solids
WAB	Waste Agricultural Biomass
WTO	World Trade Organization

Contributors

Editors

J.W.A. Langeveld is the director of a research and consulting company (Biomass Research) that focuses on the development and evaluation of bioenergy and biobased production chains. He has a background in applied agronomic and economic research. The main focus of his work is on bioenergy feedstock availability and sustainability, market analysis, and production development of biobased and decentral bioenergy production chains. Langeveld described biofuel crop production systems, studied biobased production from crops and residues, and made assessments of biomass availability. He has been a member of several scientific committees including a working group on the development of GHG emission calculation methodology (for CEN TC 383) and is involved in the actualization of the Dutch sustainability standard NEN 8080.

In the past, Langeveld has been senior researcher at Wageningen University and Research Centre and researcher at the Centre for World Food Studies in Amsterdam, where he worked on land use, nutrient management and greenhouse gas (GHG) emissions. His track record includes work on sustainable land use systems in Europe, Asia, Africa and Latin America. Hans has been leading large bioenergy projects and is (co) author of 60 scientific papers as well as books on the biobased economy and on European farming systems.

John Dixon is Principal Adviser – Research Programmes at the Australian Centre for International Agricultural Research. He has over 30 years of developing-country experience with agricultural research and development, including cropping systems, economics and natural resource management with the Consultative Group on International Agricultural Research (CGIAR) system and the Food and Agriculture Organization of the United Nations (FAO UN).

Herman van Keulen was trained as a soil scientist and production ecologist at Wageningen University. He worked in Indonesia focusing on nitrogen nutrition and modelling of rice, after which he started his career at the DLO-Centre for Agrobiological Research (CABO-DLO) in the

Department of Agrosystems Research. His main research interests were in the field of development-oriented research (e.g., the development, testing and application of crop growth simulation and optimization models).

In 1983 he was the course leader of an international modelling training course organised jointly by CABO, WMO, and FAO. He served as a guest lecturer at the Technische Hochschule in Ilmenau, Germany, was visiting professor at the International Institute for Aerospace Survey and Earth Sciences (ITC), Enschede, the Netherlands, and worked at the International Wheat and Maize Improvement Centre (CIMMYT) in Mexico. Van Keulen was professor at Wageningen Agricultural University from 1994 until he retired.

Other Contributors

Harry Croezen is a senior researcher with 20 years of experience, currently working at CE Delft, a private and independent environmental consultancy in the Netherlands. His specialisations are LCAs and economic and environmental assessments of production processes in the field of biomass cultivation and biomass processing for heat, power, transport fuels and chemicals. In this capacity he has conducted numerous studies for power companies, authorities and NGOs concerning realisation and assessment of initiatives, including cultivation of energy crops, co-firing projects, biomass CHP plants, innovative biofuel production routes, and co-processing of biomass in mineral oil refineries. He has been involved in the development of the European CEN 3838 sustainability standard and is currently involved in the actualization of the Dutch NEN 8080 sustainability standard.

Simon Delany is a geographer by training, an ornithologist by inclination, and an all-round environmental services professional. He worked for over 10 years as a Senior Technical Officer at Wetlands International and recently established an independent environmental consultancy in Wageningen, the Netherlands.

Thea Hilhorst holds an MS in tropical production systems and is a senior advisor at the Royal Tropical Institute in Amsterdam, the Netherlands. She specializes in land tenure, land governance and institutional innovation.

Keith W. Jaggard is now retired but spent 40 years doing field research into the agronomy of the sugar beet crop, most of that time at Broom's Barn Research Station in England.

M. Valentina Lasorella is a researcher at the National Institute of Agricultural Economics (INEA), working in the area of agricultural and rural development with a focus on the evaluation of the environmental impact indicators. She has contributed to research projects financed under the Med Program (6th Research FP) regarding the assessment of energy crop production and conversion techniques in the Mediterranean area and policy incentives for the evaluation of innovative features of governance regarding the renewable energy policies and projects at the local, regional, or national level with the Scuola Superiore Sant'Anna of Pisa. She is also working on projects in developing countries and aims to improve the performance and productivity of African countries, integrating land management and water resources, which are key factors to both water and food security.

Wilson José Leonardo is a PhD candidate of Plant Production Systems at Wageningen University. Wilson's research focuses on assessing the possibilities and limitations of biomass for biofuel production on smallholder farms in Central Mozambique. In 2007 he earned his master's degree in Plant Science from Wageningen University. In 2001 he was graduated as a Bachelor of Plant Production and Protection at the Eduardo Mondlane University, Mozambique. He worked for the International Crop Research Institute for the Semi-Arid Tropics from 2001 to 2008, before he joined Wageningen University for his PhD in 2009.

Emiliano Maletta is an agronomist and environmental economist with a focus on crops for noncompetitive lands. He is a researcher at CIEMAT, a public authority on energy crops in Spain and the Director of Bioenergy Crops Ltd, a consulting firm based in the UK.

P. M. Foluke Quist-Wessel is a senior agronomist and director of AgriQuest. She holds a degree in Tropical Crop Science (MS) from Wageningen Agricultural University. During long-term overseas assignments, she gained professional experience in agricultural production systems and rural development aiming for food security and agro chain development. As a researcher she worked on global food security (Plant Research International, Wageningen UR), sustainability aspects of bioenergy production and by-product valorisation (Biomass Research).

Belinda Townsend is a plant molecular biologist with interests in selecting for or modifying crops to improve their value, applications and sustainability. Dr Townsend has published in the fields of terpenoid metabolism for disease resistance in oats and oil quality in cotton, as well as defence responses in barley. Currently, she is using genomic and metabolomic approaches to study sucrose accumulation and cell wall composition in sugar beet roots to improve yield and diversify biorefining applications.

Maurits van den Berg has a background in soil science and land-use systems analysis. He earned an MS in Agricultural Sciences from Wageningen University (1986) and a PhD from Utrecht University (2000). While he is currently at the Joint Research Centre (JRC), Maurits has previously worked in the Netherlands (PBL, the Netherlands Environmental Assessment Agency, ISRIC, Wageningen University, and RIVM), Brazil (Instituto Agronomico de Campinas), Mozambique (Eduardo Mondlane University), and South Africa (SA Sugarcane Research Institute and UKZN).

1 General introduction

J.W.A. Langeveld, H. van Keulen and J. Dixon

1.1 Opening

Biofuel cropping systems are shaped by energy and agricultural policies. Following the implementation of ambitious biofuel policies during the first decade of the twenty-first century, biofuel production in many industrial and emerging countries has thrived. After Brazil had shown the way, mainly with its successful Proálcool program for ethanol, the United States of America (USA), Canada, the European Union (EU), Australia, Argentina, China, India, and many other countries developed policies to stimulate domestic biofuel production. The main justifications for the policies, and the support for biofuel producers that was part of the package, were energy security, greenhouse gas (GHG) emission reduction, rural development, and import substitution.

Introduction of the biofuel support programmes generated widespread criticism. Researchers, non-governmental organisations (NGOs), and others questioned whether the production of biofuels would really help to reduce GHG emissions and whether the price to be paid was too high. The direct costs included subsidies to farmers to produce biofuel feedstocks, as well as investment subsidies, tax exemptions, and loans, and also indirect costs such as increases in food crop prices and competition for land, water, and farm inputs.

Clearly, it is not self-evident that bioenergy is environmentally (or socio-economically) superior to fossil energy. Consumers may object to bioenergy products because of concerns about the impacts of their production. Well-to-wheel studies demonstrate that bioenergy systems vary substantially in their reliance on fossil inputs and consequently in their contribution to reduced GHG emissions (e.g., Edwards et al., 2006) – one major rationale for governments promoting these fuels and for consumers using them.

The production of renewable feedstocks can also cause negative impacts. Much attention has been directed to the possible consequences of land-use change, referring to well-documented effects of forest conversion and cropland expansion into previously uncultivated areas, with possible bio-diversity losses, GHG emissions, and degradation of soils and water bodies (e.g., Aubry et al., 2011). Sustainability concerns relating to the feedstock supply systems include direct and indirect social and economic aspects, including land-use conflicts, food security impacts, and human rights violations (Berndes and Smith, 2012).

It is surprising that there is still so much debate about biofuel production. The performance of biofuel chains has been well analysed and reported. Does this not provide sufficient information? Why is the debate so complex and why are satisfactory answers so slow in emerging? After all, biofuel production levels up until now have been quite limited, and the required volumes of feedstocks have been far less than the wastage in food chains or the amounts of food grains fed to livestock. Further, the area of land devoted to the production of feedstocks for biofuels is a small percentage of the total arable land currently in use while further arable areas remain unused.

Part of the answer may be in the quick development of biofuel production, especially in the USA and in the EU. Also, the prospects of further growth may cause concern, while the political and economic support for biofuel production might play a role as well. Finally, feared negative impacts – including the perceived contributions to the strong food-price increases during 2007–2008 – have raised awareness of the possible implications of fully developed biofuel industries.

Consequently, it is valuable to describe and analyse biofuel cropping systems worldwide. This book will do this in a balanced way. We believe there are reasons that past studies have come to such divergent conclusions. Also, we are convinced that it is important to devote time to the description of biofuel production, including the preparation of the land, the cultivation of the crops, and the conversion of biomass into biofuels. Local information (e.g., on soils, cropping systems, farm organisations) helps to explain the diversity and complexity of the conditions under which biomass is produced and converted. It can also shed light on the effectiveness of prevailing biofuel policies and the way these impact biofuel production. Most importantly, this book will help us understand how biofuel production does (or does not) fight poverty, combat hunger, and affect food production and the conservation of forests and biodiversity.

But before we do so, this chapter provides an overview of the issues that have been raised with respect to biofuels (see Section 1.2) and the lessons learned from the analysis of their production (see Section 1.3).

1.2 Points of concern

Not sustainable

In the concept of sustainability, three dimensions can be distinguished: environmental, economic, and social. In different ways, biofuels have been criticised for not being an environmentally sustainable way to reduce GHG emissions. Early studies criticising biofuel chain performance focussed on net energy production (which supposedly would be negative, requiring more energy than could be produced), low energy yields, and high costs. An excellent example of such discussions is given in the paper by Farrell et al. (2006), defending the potential role of corn ethanol in replacing fossil gasoline and reducing GHG emissions, and the array of letters following this paper (and others) in *Science* later that year.

With respect to GHG emissions, a new dimension was added to the debate when, in 2008, two papers were published on the indirect effects of biofuel production on emissions. Searchinger et al. (2008) and Fargione et al. (2008) demonstrated how changes in land requirements could provoke the conversion of uncultivated areas, thereby causing the release of large amounts of carbon stored in vegetation and soils. Both papers have been highly influential, launching a debate (on land-use change caused by biofuel production) that continues into the second decade of the twenty-first century.

The land-use debate is characterised by three elements that need to be distinguished clearly. First, one needs to determine how much land is devoted to the production of biofuel feedstocks. Second, impacts of the changes on general land use need to be assessed. Most of these effects will occur in regions at (considerable) distance from the place where biofuel crops are cultivated. Therefore, we speak of indirect effects or indirect land-use change (ILUC). Indirect effects usually refer to an expansion of agricultural area, which would be caused by the need to compensate for the loss of land that formerly was used to produce food.

The third element of the land-use debate that needs to be addressed is the amount of carbon that is released from ILUC. This depends not only on the area of land that is converted but also on the amount of carbon that was stored in vegetation and/or soil organic matter and their loss, and is particularly significant in tropical areas.

Land-use change in practice is extremely difficult to assess, as biofuel production is presently still a very small part of total global crop production (although in some regions it is significant at the local level). Consider, for example, assessments that have been made of the area that will be needed for the EU to realise its biofuel targets in 2020. Table 1.1 presents an overview of the outcomes of three modelling exercises. The area of land needed to produce biofuel feedstocks for the EU has been estimated from 0.2 to 15 million ha. On average, nearly 8 million ha might be needed in 2020. Most land will be found in South America, as large quantities of sugarcane and soybean will be imported from Brazil, Argentina, and so forth. Domestic production in the EU is expected to cover less than 30% of the extra feedstock requirements.

Crop-yield development is one of the dominant factors determining future land use. Differences between optimistic and less optimistic scenarios (assuming normal or extra yield improvement, depicted by the business as usual [BAU] and high yields [HY] scenarios of the Aglink model, respectively) clearly show the impact of yield improvement. Impacts of trade policies, depicted by two scenarios of the International Food Policy Research Institute (IFPRI) model, are smaller.

Table 1.1 Land needed to satisfy current EU biofuel targets in 2020 (million ha)

Continent	Aglink		IFPRI		Banse	Average
	BAU[1]	HY[2]	BAU	FT[3]		
EU	1.5	0.8	0.8	0.5	4.0	1.5
Africa	0.2				1.0	0.2
South America	1.7	1.0	5.2	7.3	6.0	4.2
North America	0.4	−0.01				0.1
Asia	0.6	−0.4	0.2	0.2		0.1
Russia	0.4	0.2	0.6	0.6		0.4
Oceania	0.3	0.1				0.1
Other (not specified)	0.2	−1.6	1.3	1.2	4.0	1.0
Total	5.2	0.2	8.2	9.8	15.0	7.7
EU as share of total	28%	%	10%	5%	27%	17%

[1]BAU = business-as-usual scenario
[2]HY = high yield
[3]FT = free trade
Source: Calculated from data presented by Croezen et al. (2010).

A special point of concern for biofuel feedstock production is the need for agricultural inputs such as fertilisers, water, and agro-chemicals. It is feared that substantial expansion of biofuel production may go at the expense of input availability; this could threaten food production. A major concern is the demand for phosphorus fertiliser – a finite resource that is being mined. Also, water problems may be aggravated by large-scale biofuel crop production, which may cause serious problems in areas that are already drought-prone.

Not profitable

A second point of debate on biofuels is their relative profitability. According to OECD-FAO (2009), biofuel production is increasingly driven by quantitative mandates. These either take the form of blending requirements or minimum biofuel quantities to be used in the national transport sectors. This applies not only to biofuels produced in the USA and the EU but also to countries such as Indonesia, Malaysia, and many others, including biodiesel development in Brazil. Without mandates, biofuel production is in many cases not profitable.

In many countries, the revenues from fuel excise taxes flow directly into the treasury. In some countries, however, revenues from fuel excise taxes are used to invest in transport infrastructure (OECD, 2008). Some Organisation for Economic Co-operation and Development (OECD) countries give input subsidies to enhance the use of agricultural inputs. In the EU, farmers using

fallow (set aside) land to produce bioenergy feedstocks receive a small subsidy. Other support measures include the reduction in infrastructure costs,
capital grants, guaranteed loans, capital allowance schemes, and direct subsidies per unit of biofuel produced (Steenblik, 2007).

The need for economic support to allow biofuels to compete with their
fossil counterparts has been widely criticised; for example, by the OECD
(e.g., Steenblik [2007] and OECD-FAO [2009]). Many countries have
implemented support measures, including tax credits and subsidies (e.g.,
for research). Table 1.2 lists policy support measures taken in the EU. Tax
exemptions and blending targets are the most common measures. The earli-
est measures were taken in Germany in 1999 and in Spain in 2003. In most
countries, however, biofuel support started in either 2005 or 2006.

In the USA, measures to support the production and use of biofuels
include direct subsidies, blending obligations, research grants, import tariffs, and tax exemptions for biofuel blenders. Support took off before 2006,
and an overview of measures is given in Table 1.3. Subsidies have tended
to increase since 2006, with the highest support given to the production of
first-generation ethanol.

Table 1.2 Introduction of biofuel policy instruments in selected European countries

Country	Tax Credit	Blending Obligation	Subsidy for Research and Development
Austria	2007	2005	–
Belgium	2006	2005	–
Bulgaria	2005	–	–
Czech Republic	2006	–	–
Denmark	2005	2005	–
Finland	–	–	–
France	2006	2005	2005
Germany	1999	–	–
Italy	–	–	–
Netherlands	2006	2007	2006
Poland	2004		–
Romania		2007	–
Slovakia	2004	2006	–
Spain	2003	2007	2003
Sweden	2006	–	–
UK	–	2006	–

Source: Lin and Yi (2012).

Table 1.3 Subsidies per unit of biofuels in the USA

Element	Ethanol		Biodiesel		Cellulosic Ethanol	
	Low	High	Low	High	Low	High
Subsidy per unit of liquid fossil fuel displaced in 2006 ($/GJ)	15.1	16.1	14.6	18.6	10.5	13.2
Subsidy per unit of liquid fossil fuel displaced in 2006–2012 ($/GJ)	15.3	17.0	10.8	14.7	11.1	14.6
Subsidy per unit of other fossil fuel displaced in 2006 ($/GJ)	29.3	38.9	25.8	32.9	10.1	12.7
Subsidy per unit of other fossil fuel displaced in 2006–2012 ($/GJ)	29.8	40.7	19.0	26.0	10.7	14.1
Subsidy per metric ton of CO_2 equivalent reduced in 2006 ($/ton)	na	520	No data	No data	118	147
Subsidy per metric ton of CO_2 equivalent reduced in 2006–2012 ($/ton)	na	545	No data	No data	124	164

na = not applicable

Source: Yin and Li (2012).

Table 1.4 Impact of biofuel support removal on biofuel consumption (billion litres per year)

	Brazil	USA	EU	China	Far East	World
Ethanol	−2.8	−5.3	−5.1	−1.6	No data	−16.5
Biodiesel	1.9	−0.9	−12.2	No data	−1.5	−13.1

Source: OECD (2008).

Without financial support, the consumption of biofuels would be less (Table 1.4). Elimination of import tariffs would cause relocation of ethanol production across countries, with increased exports from countries such as Brazil and higher imports in the USA, Canada, and particularly the EU. Some subsidies affect the situation in other countries. For example, biodiesel consumption in Brazil benefits from support measures taken in other countries. Removal of support would reduce biodiesel consumption in Canada by 80% (OECD, 2008).

Competition with food

Financial support for biofuel production will also affect crop feedstock prices. According to OECD-FAO (2007), biofuel production will lead to higher commodity prices, which is of particular concern for net food-importing, developing countries, as well as for the poor in urban populations, and will evoke ongoing debate on the 'food *versus* fuel' issue. Furthermore, they imply higher costs for livestock producers. The perceived role of biofuels in rising food prices has been heavily criticised.

According to OECD (2008), removal of the support for biofuels would lead to price declines for wheat, coarse grains, and oilseeds (all in the 0–10% range), and also for vegetable oils (–15%), but it would lead to increases in the price of soybean and rape meals (+8%) as well as sugar (+4%). Many consider recent (since 2007) food price increases as direct consequences of biofuel-supporting measures, but the exact contribution remains difficult to assess. According to the International Energy Agency (IEA Bioenergy, 2011), a combination of high oil prices, poor harvests, and speculation with food commodities probably had a stronger impact on food prices than biofuels. The World Bank (Baffes and Dennis, 2013) concludes that increases in fossil oil prices, changes in food crop stocks and exchange rates – and not biofuel expansion – are dominant factors explaining food crop prices since 2004.

Other impacts

Further points of concern refer to the impact that increasing land requirements may have on the position of local smallholders or other land users, especially in regions with a strong expansion of biofuel production areas. Already, incidents have been reported where the local population is suffering from industrial producers who introduce biofuel production in their regions (e.g., Van der Vlist and Heringa, 2012).

1.3 This book

The points that have been raised by critics are extremely relevant, and if biofuels want to earn their 'license to be produced', these points need to be systematically addressed. How, then, to assess the impact biofuels have on land use, input use, crop prices, and food availability? How to determine the performance of production chains around the world? From our experience in research related to crop production, we know that the sustainability of a production system cannot be determined without considering its specific environment. Cultivation of a given crop in one region – with its soils, climate, and farming systems – can take a completely different form compared to that in other regions where it is grown.

This book aims to provide a generic analysis of different biofuel production systems. It presents a technical description of existing practices around the world, providing sufficient detailed information and allowing a comprehensive, dynamic, multidisciplinary analysis of thematic issues that depict the way existing practices of land use or food production are affected by the increasing demand for biomass-based transportation fuels.

Addressing the (potential) role of biofuels in these challenges requires a comprehensive approach. It should not only include a range of crops that can serve as biofuel feedstocks, but it should also show how prevailing local conditions (soils and climate, and also farm types, infrastructure, and social organisations) codetermine how biofuel crops interact with systems that provide food, feed, fibres, shelter, or biodiversity in developing, emerging countries, as well as in developed countries.

Many studies and books tend to characterise biofuel crop production as a process decided by inherent crop characteristics that are determined by plant architecture, tissue composition, and botanical features. While this helps to explain how crops perform under given agro-ecological conditions, it tends to ignore more complex issues such as the role of biofuel crops in food or feed production, their place in crop rotations or in complex processes of land-use change, and carbon management. It also neglects the role farmers play in managing crops and realising yields.

This book analyses biofuel crop production chains by discussing practices of land use, crop cultivation, and biomass conversion. This integrated approach, which provides a more comprehensive understanding of the background of biofuel production, including its economic, environmental, and social backgrounds, is expected to better describe the performance of existing production chains, as well as to show why the chains have developed in their current forms.

The book presents an overview of major biofuel crop production systems in Brazil, the USA, the EU, Indonesia/Malaysia, China, Mozambique, and South Africa. Specific chapters will be devoted to lignocellulosic feedstocks and to the use of crop residues. Hence, this book covers all biofuel systems of global importance, plus less dominant systems of regional importance in Asia and Africa relevant in the context of today's major challenges: how to organise food and energy production while curtailing carbon emissions and land-use change.

General principles

The description and analysis of biofuel production systems presented will follow a new approach. Basically, crop and biofuel production are considered in their local context. The book explains which feedstocks are, and will be in the future, used in the production of biofuels. Special attention is given to the conditions under which production is taking place, determining common crop rotations plus practices of land preparation, input application, and harvesting. Available natural resources are presented, as are general market conditions. For each country (or a combination of countries), biofuel targets are given, plus the type – and extent – of financial support measures for ethanol and biodiesel. Common approaches for biomass conversion are presented, and yields of main and co-products are assessed.

Specific attention is given to the relation between biofuel production and the following:

- Soil type
- Land preparation practices
- Nutrient management
- Conversion technology
- Land rights.

While this is only a limited selection of the total number of factors determining the performance of biofuel production chains, these factors are essential to understanding the behaviour and performance of chains and their stakeholders. Together, they provide a sufficient overview of elements of access to land, soil conditions, farming practices, and efficiency of conversion. Thus they represent legal, edaphic, agro-ecological, economic, and technological aspects of biofuel production.

In the description of biofuel production systems, major emphasis is given to (1) land use and expansion of biofuels areas, (2) production and yield levels, (3) efficiency of input use in crop cultivation, and (4) social and economic impacts of biofuel production. The approach chosen in this book contains the following elements:

- An overview of land resources and land use based on land-use statistics providing a vital picture of land use and its dynamics over a period of 30 years
- A description of biofuel policies, including mandates and blending obligations, subsidies, and import levies (when available), plus developments in biofuel production
- Translation of enhanced biofuel production in crop feedstock requirements
- An overview of major crop production practices, including crop calendar, land preparation, input use, and crop yields
- Conversion of biomass, biofuel yields, and co-products generated
- Detailed evaluation of production chains with respect to their impact on land use, efficiency, GHG emissions, economic and social implications, and biomass availability.

Thus, the book describes entire production systems in an integrated way. It takes account of the regional setting, as defined by existing land use, policies, and crop cultivation practices, and it places generous attention on the presentation of facts, whereby the dynamic character of the processes involved is emphasised.

Public statistics are used to describe (developments in) land resources, biofuel policies, and crop production and use, as well as to assess impacts on the environment in terms of soil and water, land use, GHG emissions, economic and social effects, and food availability. Central are implications for policy development. Analyses are quantified as much as possible. Numerical values are presented mainly in tables to assure accessibility of the data and readability of the text. References are given in tables. Trends are identified and presented, and links are provided.

Organisation of the book

Central in the analysis is land use by crops that are used as biofuel feedstocks. If one wants to understand the development of the biofuel production sector and its impact on its environment, one needs to develop a thorough understanding of crop cultivation systems. Therefore, this book presents basic data on crop types, soil conditions, input use, and biomass-to-biofuel conversion. All data refer to the national (or regional) level, depicting the local situation as accurately as possible. Evaluation of the performance of different chains, however, follows a generic approach, using concepts that are generally applicable.

The book starts with the introduction of some basic concepts that are crucial in the analysis. Concepts related to land cover and land use are presented in Chapter 2. Inputs used in crop production are discussed in Chapter 3, and principles of GHG emissions are covered in Chapter 4. Chapter 5 discusses land rights and how those can be affected by biofuel production.

Biofuel production chains are then presented in detail. Those include the production of sugarcane-based ethanol and biodiesel from soybeans in Brazil (Chapter 6), corn ethanol and soybean biodiesel in the USA (Chapter 7), wheat- and rapeseed-based biofuel production in the EU (Chapter 8), sugar beet ethanol in the EU (Chapter 9), palm oil biodiesel in the Far East (Chapter 10), and ethanol production in southern Africa (Chapter 11) and China (Chapter 12). Further chapters present alternative biofuel feedstocks: lignocellulosic crops (Chapter 13) and agricultural wastes and residues (Chapter 14). The final two chapters present an overview of the results from the regional chapters (Chapter 15) and conclusions (Chapter 16).

References

Aubry, S., Seufert, Ph., and Monsalve Suárez, S. (2011). *(Bio)fuelling injustice? Europe's responsibility to counter climate change without provoking land grabbing and compounding food insecurity in Africa*. Rome, Italy: Terra Nuova.

Baffes, J., and Dennis, A. (2013). Long term drivers of food prices. Policy Research Working Paper 6455. Washington, DC, USA: World Bank.

Berndes, G., and Smith, T. (2013). Biomass feedstocks for energy markets. In J. Tustin (Ed.), *2012 Annual Report*. Paris, France: International Energy Agency.

Croezen, H.J., Bergsma, G.C., Otten, M.B.J. and van Valkengoed M.P.J. (2010) *Biofuels: indirect land use change and climate impact*. Delft, the Netherlands: CE Delft.

Edwards, R., Larivé, J.-F., Mahieu, V., and Nouveirolles, P. (2006). Wells-to-wheels analysis of future automotive fuels and powertrains in the European context. Well-to-tank report. Version 2b. Concawe, Eucar, Joint Research Centre. Ispra, Italy: Institute for Environment and Sustainability of the EU Commission's Joint Research Centre.

Fargione, J., Hill, J., Tilman, D., Polasky, S., and Hawthorne, P. (2008). Land clearing and the biofuel carbon debt. Supporting material. *Science,* Vol 319, pp1235–1238.

Farrell, A.E., Plevin, R., Turner, B.T., Jones, A.D., O'Hare, M., and Kammen, D.M. (2006). Ethanol can contribute to energy and environmental goals. *Science,* Vol 311, p506. (Letters following this paper: *Science,* Vol 312 (23 June 2006).)

IEA Bioenergy. (2011). *Technology roadmap: Biofuels for transport*. Paris, France: International Energy Agency.

OECD. (2008). *Economic assessment of biofuel support policies*. Paris, France: Organization for Economic Cooperation and Development.

OECD-FAO (2007) *Agricultural outlook 2007-2016*. Paris, Organization for Economic Cooperation and Development; Rome, Food and Agricultural Organization of the United Nations.

OECD-FAO. (2009). *Agricultural outlook 2009–2018*. Paris, France: Organization for Economic Cooperation and Development; Rome, Italy: Food and Agricultural Organization of the United Nations.

Searchinger, T., Heimlich, R., Houghton, R.A., Dong, F., Fabiosa, J., Tokgoz, S., Hayes, D., and Yu, H.-T. (2008). Use of U.S. croplands for biofuels increases greenhouse gases through emissions from land-use change. *Science,* Vol 319, pp1238–1240.

Steenblik, R. (2007). *Subsidies: The distorted economics of biofuels*. Geneva, Switzerland: The Global Subsidies Initiative, International Institute for Sustainable Development.

Van der Vlist, L., and Heringa, S. (2012). *Impacts of the Dutch economy on indigenous peoples*. Amsterdam, the Netherlands: Netherlands Centre for Indigenous Peoples.

Lin, C.-Y., and Yi, F. (2012). *What factors affect the decision to invest in a fuel ethanol plant? A structural model of the investment timing game in Europe*. Retrieved 6 March 2013 from http://www.osti.gov/eprints/topicpages/0460/binge+ethanol+evidence.html

2 Land cover and land use

J.W.A. Langeveld

2.1 Introduction

Concerns referring to sustainability, profitability, and impact of biofuel production on food availability cited in Chapter 1 are not unique. Historically, warnings have been given about the impacts of ever-increasing demands for food and natural resources. In 1972 a group of researchers at the Massachusetts Institute of Technology (MIT), headed by Donella Meadows, wrote the probably most influential publication about the limitations of the earth's ability to generate food and biomass. *The Limits to Growth* was commissioned by a think tank called the Club of Rome. Findings in the book were based on calculations with a computer model. Including only a few variables, the model assumed exponential demand for food and products while technology and resource use could grow only linearly. Population growth, agricultural productivity, and environmental protection were among the key issues that were studied (Meadows et al., 1972).

The outcome of Meadows's study was that unbalanced economic and population growth could not be supported by existing finite resource supplies. In reaching that conclusion, the book echoed older concerns and predictions, such as those reached by Thomas Malthus in an essay on population growth published in 1798. Both share a concern that agriculture may be unable to produce the crops needed to feed a population that is not only growing but is also using more biomass per head each year. Declining availability of arable land per person is often cited as an indicator that we are losing agricultural production potential. The finite pool of natural resources – land, water, and genetic potential – would no longer be sufficient to cover the biomass requirements compounded by competing demands from urbanisation, industrial uses, and biofuel production. At the same time, crop productivity currently is threatened by processes such as climate change (Nachtergaele et al., 2011).

But others are not so pessimistic. Lambin (2012) identified two competing views on land availability. On the one side, the Malthusian view emphasises that the stock of suitable land is limited. Increasing demand for productive land will evidently lead to competition between alternative types of land use. Eventually, a shortage of productive land will have negative impact on welfare. The alternative way of thinking, referred to as the Ricardian view, suggests that increasing demand for land will make it economically feasible to start cultivating marginal land as prices of land-based commodities increase. A geographic redistribution of land use, trade, and investments in land resources will provide access to more resources worldwide, be it at increasing economic, environmental, and social costs. In this view, global food security will increasingly involve a trade-off between food and nature (Lambin, 2012).

The Food and Agricultural Organization (FAO) of the United Nations has a long tradition of assessing the availability of natural resources such as land and water in the light of increasing demands. In a series of studies, projections have been made on how these demands are to be fulfilled. A good example is given by Nachtergaele et al. (2011). Their analysis suggests that there are large areas of presently unused arable land, most of which is found in developing countries (a figure of 2.6 billion ha is presented, with 1.8 billion in developing countries).

This study also shows, however, that area expansion is *not* to be the main source of additional biomass. Most increases in crop production are expected to be generated by intensification. In developing countries, yield increases (71%) and higher cropping intensities (8%) will generate 80% of enhanced output levels. This is an average value; the role of intensification locally can rise to 95% in land-scarce regions (South Asia, the Near East, and North Africa). According to the authors, arable land expansion will remain an important factor in less constrained regions like sub-Saharan Africa and Latin America. But even here, expansion will be less important than it was in the past.

The precise outcome of expansion and intensification will be determined by relative prices of crops, land, labour, and other agricultural inputs, but it is projected that intensification (higher yields and more intensive use of land) will contribute more than 90% to growth in crop production at the world level up to 2050 (Nachtergaele et al., 2011).

This conclusion provides a stark contrast to projections that have been cited on the impact of biofuel production on food availability. It would be interesting to see to what extent expansion or intensification has played a role in supporting the increasing biofuel production since the turn of the twenty-first century. To do so, we need to introduce a few basic concepts related to land use and biomass production.

First, it is of crucial importance to make a distinction between land *cover* and land *use*. In discussions on land availability, these terms are often confused. Land cover refers to the ecological state and physical appearance of the land surface based on a classification system (e.g., forests, grasslands, or savannahs) (Turner and Meyer, 1994). Change in land cover reflects a shift based on a defined classification, regardless of land use. Changes in land-cover classification can result from how data are interpreted or aggregated, the scale and order of analysis, and also from actual physical changes that cross the threshold values that define a given land-cover class (Dale and Kline, 2013).

Referring to changes projected previously by Nachtergaele et al., area expansion relates to changes in land cover while intensification refers to (changes in) land use. Contrasting views on land use and its impact on biomass availability – affecting food provision and leading to deforestation or opening up of virgin nature areas – will play an important role in the background of biofuel production routes discussed in this book.

This chapter discusses major concepts in land cover and use, including the following elements: Section 2.2, total land surface and different types of land use; Sections 2.3, aspects of climate and soil quality; and Section 2.4, how trends in land cover and land use can be extracted from statistical records.

2.2 Land cover and land use

Following the way land-use data have been organised in the FAO Statistical Database (FAOSTAT), this book makes use of the following equations:

$$\text{land area} = \text{forest} + \text{agricultural land} + \text{other land} + \text{inland water} \tag{2.1}$$

$$\text{agricultural land} = \text{permanent grassland} + \text{agricultural tree crops} + \text{arable area} \tag{2.2}$$

These equations allow for the composition of detailed land-use balances of any country covered in FAOSTAT. Note that Equation 2.1 relates to main land-cover types, whereas Equation 2.2 refers to land use.

The FAO defines the following types of land cover: forest, agricultural land, inland water, and 'other land'. The latter covers all non-forest, non-agricultural cover types including urban areas, infrastructure, and touristic areas, as well as deserts, mountainous areas, permanent ice covers, and probably also abandoned (ex-agricultural) land. *Forest* refers to land with a permanent forest cover. *Agricultural land* includes permanent grassland, arable land, and other permanent crops. It may be idle, cultivated, or under fallow.

According to the FAO, the Earth's surface contains 13.5 billion ha of continental area, with 0.5 billion consisting of inland water and some 13 billion consisting of land. Of these, most land (4.9 billion ha) is classified as agricultural. Forest and other land types each cover about 4 billion ha (Table 2.1).

Table 2.1 Area covered by main land cover types (billion ha)

	Area Covered (billion ha)	Definition
Forest land	4.0	Permanent forests for non-agricultural use. Includes non-natural forests for commercial use. Distinction with forested agricultural land may be difficult to see.
Agricultural land	4.9	Mostly covered by food, feed, or industrial crops, mainly in use for agricultural or industrial purposes. Distinction between extensively used agricultural land and non-agricultural land (forest or other land) is sometimes difficult to make.
Other land	4.1	Cities, infrastructure, other industrial areas (e.g., mines), nature areas, permanent ice, mountainous areas, deserts. Distinction with extensively used agricultural land may be difficult to see, especially when using remote sensing.
Inland water	0.5	Lakes, inland seas.
Total land surface	**13.5**	

Source: FAOSTAT (2010-2013); http://www.faostat.fao.org.

It is important to develop a good understanding of the way agricultural land is used. Rewriting Equation 2.2 gives:

$$\text{agricultural tree crops} = \text{agricultural land} - \text{permanent grassland} - \text{arable area} \quad (2.3)$$

Further, arable area is defined as:

$$\text{arable area} = \text{arable crop area} + \text{fodder crop area} + \text{fallow area} \quad (2.4)$$

Fallow area is arable land that is not currently cropped but has been cropped in the past and probably will be cropped again in the future. While it is not actively cropped, in practice, fallow land is rarely idle. It can be used, for example, as grazing land and for producing firewood, honey, and thatching material. This is further discussed in Chapter 5.

The classical view on intensification of cropping systems (as defined by Ruthenberg (1980) is that a higher cumulative output per ha, and therefore a higher human 'carrying capacity', can be obtained by increasing the cropping intensity. In this light, fallow rotation systems are considered an intermediate stage between 'shifting cultivation' or 'long-rotation fallow systems', where land is cropped less than one-third of the time, and 'continuous cropping', where land is cropped more than two-thirds of the time (van Noordwijk, 1999). In current agricultural systems, shifting cultivation is very rare, and most farmers practice continuous cropping. Equation 2.3 is used to calculate the area of agricultural tree crops. Equation 2.4 can be used to calculate fallow area. This book strives to present complete land balances for each region or country discussed, as this will provide a better insight into the background of recent land-use changes. In practice, however, data on fodder crops and fallow often are incomplete.

The book presents graphics on land cover and land use, which allows readers to get a good (quick) assessment of the way land is used in a given area. Consider, for example, Figure 2.1, which depicts the agricultural land use defined in Equation 2.2 (Figure 2.1a depicts arable, grassland, and tree crop areas). Permanent grassland is the dominant land-use type, making up more than two-thirds of all agricultural land. Arable area covers 28%, with tree crops globally representing 3% of all agricultural land (Figure 2.1b depicts total global land use). This picture clearly demonstrates the fact that non-agricultural land use dominates over agriculture, while – within agricultural land use – the area of arable cropland is relatively small (although it is larger than fodder, fallow, and agricultural tree crops taken together). It is stressed that estimates of fodder crop and fallow area generally are very unreliable.

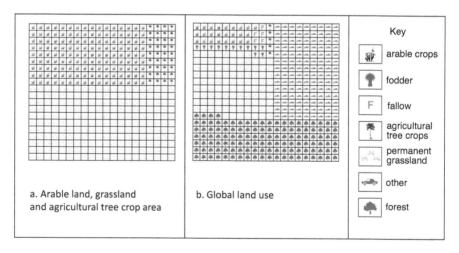

Figure 2.1 Use of land: agricultural (a) and global (b)

Sources: FAOSTAT (2010–2013); http://www.faostat.fao.org.; and Siebert et al. (2010).

2.3 Climate and soil quality

The types of land use and yields that can be realised are largely determined by climatic and soil conditions. *Climate* encompasses long-term information on temperature, humidity, wind, precipitation, and so on. It should not be confused with weather, which is the present state in a given location plus variation over relatively short periods. Climate classification systems provide an overview of regions with similar climatic conditions. One system was suggested in 1884 by the Russian-German climatologist Wladimir Köppen. The system is based on the concept that native vegetation is the best expression of climate conditions. Climate zone boundaries, therefore, largely coincide with vegetation types (see Annex Table A.1 at the end of this chapter). The Köppen system combines average annual and monthly temperatures and information on the distribution of precipitation.

The main characteristics of climatic zones relevant to the biofuel production systems presented in this book are located in tropical savannah and tropical monsoon (Brazil), tropical rainforest (Far East), temperate (USA, EU, southern Africa), and arid (western USA, southern EU, southern Africa) zones (Figure 2.2). China knows many climate types, varying from tropical to cold and arid.

Soils are crucial for profitable and sustainable crop production. They provide nutrients and water and give support. There are several systems for soil classification, and the FAO has provided a generic classification system that was used for a global soil map (Figure 2.3). Soils in Latin America are dominated by weathered clay soils in the northeast (Brazil) and more productive soils in the south. Soils in North America are mostly productive clay soils with coniferous or borest forest (spodic) soils and peat soils in the north and poor sandy and clay soils in the east. Asia's soil distribution is similar to that of North America: predominantly peats and spodic soils in the north, less fertile clays in the southeast, and large areas of productive clays and sandy soils in the remainder of the continent.

Soils in Africa are mostly weathered clay soils in the north and south, with a zone of wetland soils running in the middle of the continent from the Sahel to Egypt and large parts of the southeast. Heavily weathered ferralsols dominate large parts of West and Central Africa. An overview of the FAO and US Department of Agriculture (USDA) soil types is given at the end of the chapter in Annex Table A.2.

Major problems in soils include poor nutrient status, physical limitations, damage due to soil erosion, and chemical restrictions. An overview of soils affected is given by Fischer et al. (2008). Problems with nutrient limitations are predominantly found in Brazil, the southeastern part of the USA, northern Europe, southeast China, and on different locations in southern Africa. Poor rooting conditions and physical limitations for soil preparation are found in Brazil, the western part of the USA, southern

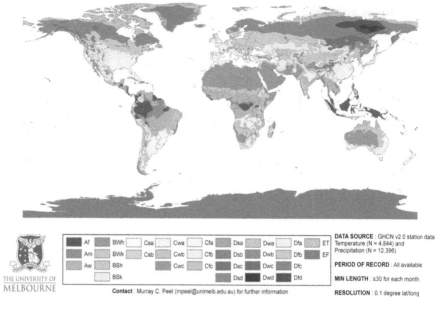

Figure 2.2 Main climatic zones of the world
See Annex Table A.1 for an explanation of Köppen's climate symbols.
Source: Peel et al. (2007).

Europe, and southern Africa, as well in southern China. The exact impact of these limitations will depend on local conditions, where multiple restrictions for crop growth may provide add-up. Some restrictions may be overcome, but usually options for soil improvement are limited while costs often are prohibitive.

2.4 Consequences for crop cultivation

The main impact of climate on crop cultivation is through limitation of the growing period – the time a crop can grow in the field. This period is limited by temperature in temperate and (sub)arctic regions (and at higher altitudes in the tropics) and by water availability in most of the other regions. Agriculture in temperate regions follows an annual cycle, usually allowing only one major crop to be grown on a piece of land in a given year. In large parts of the tropical zones, however, water and temperature conditions are such that they can facilitate cultivation of more than one crop. Cultivation of two subsequent crops on the same plot is referred to as *double cropping*. Both crops should be allowed to complete a full growth cycle, thus generating two complete harvests in one calendar year.

Double cropping is not to be confused with *sequential cropping*, where one crop is sown under the previous crop. Usually, this is done to make use of valuable nutrients and water that is available after harvest of the first crop. Sequential cropping is also used to reduce loss of residual nutrients as a way to improve quality of ground or surface water.

Double cropping (or the more intensive triple cropping) practices are an extremely powerful way to make optimal use of high-quality land in regions with favourable climatic conditions and high population density. In these regions, harvested crop area tends to exceed the amount of arable land. Following Beets (1982), the Multiple Cropping Index (MCI) is defined as the ratio of area harvested and arable crop area in a given region or country (Equation 2.5):

$$\text{MCI} = \text{harvested arable crop area} / \text{arable area} \qquad (2.5)$$

MCI is dimensionless. While it can reach values greater than 2.0, in practice it is about 1.5 in densely populated tropical areas and less than 1.0 in temperate countries. MCI is the end result of many processes and decisions taken by farmers, but it is also influenced by weather conditions. In years when rainfall starts later than normal, farmers will postpone sowing and planting. In some cases, this delay may result in a reduction of crop areas. In years with heavy rainfall, farmers may encounter problems in sowing or during harvesting. This may also affect MCI.

MCI is not a constant but changes over time. Table 2.2 shows how MCI for global land use increased from 0.85 in 1980 to 0.99 in 2010. During that period, harvested crop area increased by about 230 million ha while arable land area increased by only 36 million ha. In industrial countries,

Table 2.2 Global land area and land use (in million ha)

	1980	*1990*	*2000*	*2010*
Forest area	No data	4,168	4,085	4,033
Other land	No data	4,038	3,992	4,088
Agricultural area	4,666	4,858	4,942	4,894
Permanent grassland	3,213	3,335	3,425	3,353
Agricultural tree crops	No data	116	132	152
Arable land	1,352	1,404	1,385	1,388
Harvested crop area	1,145	1,303	1,287	1,374
Multiple Cropping Index	0.85	0.92	0.93	0.99

Source: Elsewhere this is given as FAOSTAT (2010-2013); http://www.faostat.fao.org

agricultural and arable land is declining but harvested crop area tends to remain more or less constant. This development – declining arable crop area combined with increased harvesting intensity – is typical for many regions in the world. It helps to explain why shrinking agricultural areas can support increasing populations.

Table 2.2 also shows that the area of arable land started to decline after 1990, while grassland decline set in 10 years later. Loss of forest area is increasing – since 2000, more than 70 million ha have been lost.

2.5 Conclusion

While there is huge debate on the impact biofuels may have on production of biomass for food, feed, and fibre markets, there are different views on the potential production levels that can be supported by existing land and other resources. This chapter reviewed how both land expansion and intensification can contribute to enhanced biomass production. Basic concepts related to land cover and land use were introduced. Data on land cover and use are sometimes confusing, and this chapter explained how a land balance is calculated. Overviews of climates and main soil types were given. A global land balance was presented that shows that use of arable land is intensifying. By making more effective use of available arable land, more biomass is made from shrinking arable land areas. Other land use and agricultural tree crop areas have grown, and this has mainly happened at the expense of forest land.

References

Beets, W. C. (1982). *Multiple cropping and tropical farming systems*. Aldershot, UK: Gower.

Dale, V. H. and Kline, K. L. (2013). Issues in using landscape indicators to assess land changes. *Ecological Indicators,* Vol 28, pp91–99.

FAO-UNESCO. (2007). Digital soil map of the world. Retrieved 27 February 2013 from http://www.fao.org/geonetwork/srv/en/metadata.show?id=14116

Fischer, G., Nachtergaele, F., Prieler, S., van Velthuizen, H. T., Verelst, L., and Wiberg, D. (2008). *Global agro-ecological zones assessment for agriculture*. Laxenburg, Austria: IIASA; Rome, Italy: Food and Agricultural Organization.

Lambin, E. F. (2012). Global land availability: Malthus versus Ricardo. *Global Food Security,* Vol 1, pp83–87.

Meadows, D. H., Meadows, D. L., Randers, J., and Behrens, W. W. (1972). *The limits to growth.* New York: Universe Books.

Nachtergaele, F., Bruinsma, J., Valbo-Jorgensen, J., and Bartley, D. (2011). *Anticipated trends in the use of global land and water resources*. Rome, Italy: Food and Agricultural Organization of the United Nations.

Norman, M.J.T., Pearson, C. J., and Searle, P.G.E. (1995). *The ecology of tropical food crops*. Cambridge, UK: Cambridge University Press.

Peel, M. C., Finlayson, B. L., and McMahon, T. A. (2007). Updated world map of the Köppen–Geiger climate classification. *Hydrology and Earth System Sciences,* Vol 11, pp1633–1644.

Ruthenberg, H. (1980). Farming systems in the tropics. Oxford, UK, Clarendon Press.

Siebert, S., Portmann, F. T., and Döll, P. (2010). Global patterns of cropland use intensity. *Remote Sensing,* Vol 2, pp1625–1643.

Turner, B. L. and Meyer, W. B. (1994). Global Land-Use and Land-Cover Change: An Overview. In: W. B. Meyer and B. L. Turner (eds), Changes in Land Use and Land Cover: A Global Perspective. Cambridge University Press, Great Britain, pp. 1–9.

van Noordwijk, M. (1999). Productivity of intensified crop-fallow rotations in the Trenbath model. *Agroforestry Systems,* Vol 47, pp223–237.

Annex

Table A.1 Description of Köppen's climate symbols

Main Climate Type	Sublevel	Second Sublevel
A = Tropical	f = rainforest	
	m = monsoon	
	w = savannah	
B = Arid	W = desert	
	S = steppe	h = hot
		c = cold
C = Temperate	s = dry summer	
	w = dry winter	
	f = no dry season	a = hot summer
		b = warm summer
		c = cold summer
D = Cold	s = dry summer	
	w = dry winter	
	f = no dry season	a = hot summer
		b = warm summer
		c = cold summer
E = Polar	t = tundra	
	f = frost	

Source: Peel et al. (2007).

Table A.2 Major soil types under FAO and USDA classification systems

USDA	FAO	Remarks
Oxisols	Ferralsols, gleysols	Weathered clay soils, heavy clay soils
Ultisols	Acrisols, nitosols	
Entisols	Fluvisols, lithosols, gleysols, regosols, arenosols	Young soils
Alfisols	Luvisols, planosols, nitosols	
Inceptisols	Cambisols, gleysols, solonchaks, rankers	
Vertisols	Vertisols	Clay soils
Aridisols	Yermosols, xerosols, solonetz, solonchaks	
Mollisols	Chernozems, phaeozems, kastanozems, greyzems, rendzinas, gleysols, solonetz, solonchaks	
Andisols	Andosols	Volcanic soils
Histosols	Histosols	
Spodosols	Podsols	Coniferous, boreal, or eucalypt forest soils

Source: Norman et al. (1995).

3 Input use and crop production

*J.W.A. Langeveld, H. van Keulen
and S. Delany*

3.1 Introduction

Following ambitious biofuel policies, a strong increase of feedstock production has been projected in many parts of the world. As was indicated in earlier chapters, this has raised concerns about the need for crop inputs. While there is no theoretical maximum of nitrogen that is available, water and phosphorus use is limited. Water availability is determined by local weather conditions, which cannot be controlled, and agriculture is already the largest consumer of freshwater resources in many regions. Phosphorus is mined from a very limited number of locations.

Many studies have addressed the potential threat of large-scale input use on behalf of biofuel feedstocks. This book does not try to draw generic conclusions on the way input availability will be affected by expansion of biofuel production. Instead, this text intends to explain how production chains differ with respect to their need for (scarce) inputs and how an efficient organisation of the production chain may affect the efficiency of input use. This requires a slightly broader approach than normally found in biofuel literature. The book does, for example, give an introduction to basic principles of crop production (photosynthesis, role of nutrients) to explain why differences exist in photosynthetic efficiency between crops and how nutrient inputs affect crop growth. It also introduces ways to calculate input-use efficiency.

This chapter further presents three concepts that are used throughout the book. First, it is explained how input use affects the energy balance of biofuel production chains; this information will be used in the chapter on GHG emission calculations (Chapter 4). Next, the concept of a yield gap is introduced, explaining how it can depict room for yield improvement – information that will be used in reflections on future yield development of biofuel feedstocks. The third concept that is introduced relates to the calculation of biomass production per unit of land, which is relevant for the assessment of the impact of biofuel expansion on food availability.

The remainder of this chapter is organised as follows: photosynthetic pathways of biofuel crops are discussed in Section 3.2, followed by a description of major agricultural inputs in Section 3.3 and an explanation in Section 3.4 of how energy use in biofuel production chains is affected by input use. Further sections discuss concepts of crop yields (Section 3.5) and ways to calculate crop production levels (Section 3.6).

3.2 Photosynthetic efficiency

Photosynthesis is a metabolic process in which solar energy is used to convert carbon dioxide (CO_2) and water into sugars and oxygen (O_2). The amount of energy captured by photosynthesis is immense and exceeds our energy demand manifold. The importance of light interception and photosynthetic pathways cannot be overstated, because they provide the basic energy for life on Earth. Over time, we have developed a good understanding of the chemical processes that are used to catch light energy and convert it into biomass.

Basically, photosynthesis is a two-stage process. In the first step, the *light reaction*, light energy is stored in molecules that serve as energy carriers (ATP and NADPH) in plant cells. Subsequently, this energy is used to form carbohydrates from CO_2, the *dark reaction*. The dark reaction follows one of two possible biochemical pathways characterised by the length of the C-skeleton of the first stable product, which can be three or four C-atoms. The pathway is specific for plant families; plants can therefore be denoted as either C3 or C4 plants.

The C3 photosynthetic process takes place in the chloroplasts of leaf cells. The first CO_2 acceptor, rubisco, also reacts with oxygen and has to be recovered for photosynthesis. This energy-consuming process is called *photorespiration*. Under normal conditions (340 ppm CO_2 and 21% O_2), photorespiration reduces gross photosynthetic efficiency by about one-third. The reduction is larger under O_2-rich and low CO_2 conditions and smaller under contrasting conditions (which is the case at increased CO_2 levels) that also lead to climate change.

C4 plants have two types of chloroplasts in mesophyll cells and in parenchymatic cells in the bundle sheaths, respectively. In the mesophyll, PEP (phosphoenolpyruvate)-carboxylase is the primary CO_2 acceptor. PEP-carboxylase has a higher affinity for CO_2 than rubisco and no affinity for oxygen. The fixed CO_2 is transported to the parenchymatic cells and released there. The resulting higher CO_2 level supports the functioning of rubisco in CO_2 entering the Calvin Cycle. This two-step process, which reduces photorespiration to a very low level, has a higher overall efficiency than the C3 pathway (see, e.g., Langeveld and van de Ven, 2010).

The activity of enzymes involved in photosynthesis is temperature dependent. The optimum temperature for the C3 pathway is lower (20°C) than that for C4 crops (35°C), which makes C3 crops (e.g., wheat, potatoes) more suited for cultivation in temperate regions and C4 plants (corn, sugarcane) better adapted to (sub)tropical regions. C4 plants are also more efficient in

Table 3.1 Main characteristics of C3 and C4 pathways

	Unit	C3 Plants	C4 Plants
Maximum photosynthetic rate[1] Grasses	CO_2 uptake/ $\mu mol\ mm^{-2}\ s^{-1}$	5–15	30–60
Crops	CO_2 uptake/ $\mu mol\ mm^{-2}\ s^{-1}$	20–40	30–60
Optimum temperature	°C	15–25	30–45
Photosynthetic capacity	kg CO_2/ha/h	40	70
Nutrient-use efficiency		Modest	High
Water-use efficiency	g biomass/kg water	1–5	3–5
Major crops		Temperate cereals (wheat, barley), pulses, oil crops (soybean, rapeseed), palms (oil palm)	Tropical cereals (corn, sorghum), tropical grasses (sugar cane)

[1]Maximum net photosynthesis measured at natural levels of CO_2 availability, saturated light intensity, optimum temperature, and adequate water.

Sources: Bonan (2002); Norman et al. (1995); 'C$_4$ Carbon Fixation' (2013).

water use. At a temperature of 30°C, C4 grasses lose 277 water molecules per CO_2 molecule that is fixed as compared to 833 water molecules lost by C3 grasses ('C$_4$ Carbon Fixation', 2013). As a result, they produce more biomass per unit of water used.

An overview of C3 and C4 pathways is presented in Table 3.1. C4 plants can make more efficient use of sunlight, realising a higher maximum photosynthetic rate. They thrive under conditions of high light intensity and high temperatures, and they make more efficient use of nutrients and water. Main examples of C4 crops are tropical cereals such as corn and sorghum and tropical grasses such as sugarcane. Plants using C4 photosynthetic pathways are relatively scarce, representing only 5% of all plants and 3% of all plant species. They do, however, account for one-third of all plant carbon fixation ('C$_4$ Carbon Fixation', 2013).

3.3 Crop inputs

Plant nutrients are essential for plant growth and maintenance. Nitrogen, phosphorus, sulphur, potassium, calcium, and magnesium are of special importance. They are referred to as *macronutrients* because they are required in large quantities. *Micronutrients*, which are essential but required in smaller quantities, include iron and zinc. Nitrogen and sulphur are mostly involved in enzymatic processes and protein formation. Phosphorus is part of essential esters and plays an essential role in energy transfer; nutrients such as potassium and calcium help to stabilise membranes and establish osmotic potentials. The main focus in crop fertilisation is on nitrogen, phosphorus, and potassium; their availability is essential for undisturbed growth and optimal production.

Nitrogen is an indispensable constituent of essential organic compounds such as amino acids, proteins, and nucleic acids (Mengel and Kirkby, 1987). Sufficient nitrogen is essential for undisturbed growth and development of the plant. Plant nitrogen consists of both inorganic and organic forms. Under field conditions, inorganic nitrogen mostly consists of nitrate, which can be assimilated in amino acids or stored in lower stem parts. Amino acids, formed in leaf material, may be used to synthesise proteins. Most amino acids are stored temporarily as proteins until they are needed, for example, in seed filling (Schrader, 1984). This makes nitrogen an important nutrient for the development of ears and grains of cereal crops (Mengel and Kirkby, 1987).

Minimum protein content in the vegetative phase in C3 plants is about 6% of the dry matter. The leaves of mature C4 grasses, grown with a marginal supply of nitrogen, contain only 3% proteins. This difference in protein content is related to the enzymes involved in photosynthesis. Because conversion of sugars to proteins requires energy, the low protein content of C4 crops adds to their photosynthetic efficiency. Consequently, C4 crops require less nitrogen during growth (Langeveld and van de Ven, 2010).

Plant phosphorus is mostly found in the nucleus, chloroplasts, mitochondria, cytoplasm, and vacuole (Mengel and Kirkby, 1987). It occurs in organic and inorganic forms, both forms including essential compounds that play a central role in the metabolism of the plant. The role of inorganic phosphorus depends on its location (vacuole phosphate, e.g., constituting reserve material and influencing enzyme activity; chloroplast phosphate controlling selection of photosynthetic products). The role of organic phosphorus compounds depends mainly on their form (phosphorylated sugars and alcohols, e.g., serving as intermediary compounds of metabolism, while phosphated lipophilic compounds play an essential role in membranes) (Mengel and Kirkby, 1987). Triphosphates, by far the most important group of organic compounds, include ATP, a co-enzyme that plays a unique role in the energy transport of the plant.

Water accounts for 70–90% of the fresh weight of non-woody plant species and for 50% in woody species. Water is mainly found within cells, providing a medium for biochemical reactions. It also serves as a means of transport for nutrients and photosynthetic products within the plant while maintaining turgidity (water-based stiffness) that is essential for growth and development. The amount of water involved in these processes accounts for only 5% of the water use by a crop as most water is lost via transpiration through the stomata of leaf cells (Langeveld and van de Ven, 2010).

Transpiration water lost from the stomata is replenished by uptake via the roots from the soil. The soil system is replenished through precipitation, irrigation, and capillary rise from groundwater. As long as the supply covers the demand, nutrient transport, along with the water flow from roots to the leaves, is secured.

Input use efficiency

This book quantifies the use of agricultural inputs. When possible, use of fertilisers and agro-chemicals will be presented for biofuel crop production in the different countries included in the analysis. Because there is concern about the large quantities of inputs that are associated with biofuel feedstock cultivation, efficiency of input use will be evaluated. This can be expressed in different ways.

The simplest approach would be to evaluate input use per unit of land (e.g., kg of fertiliser nutrient per ha). However, this does not reflect differences in crop yields or in soil properties. Alternatively, input use per kg of crop would not reflect variations in dry-matter content. Yield can be expressed in fresh weight or in dry-matter weight, and the moisture content of crop products shows large variations. It may be 15% for grains and 80% for potato tubers. Expressing input use per kg of crop dry matter could overcome this problem, but that will not reflect differences in efficiencies of biomass-to-biofuel conversion.

Expressing input use per unit of energy (e.g., kg of fertiliser nutrient per GJ) does not reflect differences in input-use intensity. Impacts of input use depend, to a large extent, on the amount applied per unit of land, while this tends to depend on nonlinear processes (damage often increases dramatically after certain threshold values have been surpassed). To conclude, there is not one single approach that can be used in the evaluation of biofuel crop production. Therefore, input use is expressed in a number of ways that seem the most appropriate.

Fertiliser use is calculated in two ways. Crop output – expressed in kg of dry matter – is calculated per kg of fertiliser nitrogen that is applied. Further, the book provides the amount of biofuel energy that is produced per kg of fertiliser nutrient (both nitrogen and phosphorus). Application of agro-chemicals is expressed in kg of active ingredient (active ingredients = chemical compound that affects weeds, insects, or micro-organisms) per ha.

While good progress has been made recently in evaluating and calculating the effects on the water cycle of growing biofuel feedstocks (e.g., Fingerman et al., 2011; Langeveld et al., 2012), the complexity and variability of many of the parameters that are included in these evaluations does limit their practical value. The complexity is compounded by the overlapping nature of many of the parameters and by the different ways they may be expressed at given latitudes, climate regimes, landscape types, and agricultural scenarios.

As a result, practitioners on the ground often have difficulty applying the suggested evaluations and calculations in a meaningful way. Yeh et al. (2011) adopted a practical approach in their demonstration of accounting methods for calculating water demands in bioenergy production. Their case studies at different scales provide practical examples, but many more such examples are required to meet the need for a realistic range of methodologies with practical value in the real world.

Issues that need extra attention include the relation between water consumption and general layout of the cropping system; that is, soils cultivated, land preparation, and crop management under the prevailing weather conditions. Further, downstream chain performance needs to be considered in the analysis. This is especially relevant where water consumption is expressed per unit of fuel produced (e.g., Mekonnen and Hoekstra, 2010; Yeh et al., 2011). Variations in biomass-to-biofuel conversion efficiency and of biofuel plant production processes may play a large but often overlooked role in determining the outcome of water consumption analyses. Finally, landscape elements and the regional setting of the crop production units will have a large impact, both on the consumption of water by the crop and on the impact this may have on its environment.

It is beyond the scope of this book to analyse input (fertilisers, water, agrochemicals) use in great detail. Instead, available information on application levels as well as assessments of the impacts will be presented in a consistent and transparent way. While no meta-analysis will be done in this respect, the overview will be used to discuss perspectives and implications of enhanced biofuel production in different parts of the world.

Examples of the efficiency indicators used in this book are given in Table 3.2, which also provides typical values for both C3 and C4 crop types. Data presented in this table demonstrate that (1) C4 crops are more efficient users of nitrogen or water, (2) efficiency of biofuel chains shows

Table 3.2 Indicators for efficient input use

Indicator	Unit	Typical Values	
		C3 Crops	C4 Crops
Nitrogen productivity (crop)	kg dry matter/kg of nitrogen[1]	110–140	150–200
Nutrient productivity (biofuel)	GJ/kg of nitrogen applied[1]	0.3–2.0	0.7–2.1
	GJ/kg of phosphate applied	0.5–1.5	1.0–2.0
Water use (crop)	kg water/kg dry matter	250–400	150–300
Water requirement (biofuel)	m³/GJ	35–120	20–150

[1]Values do not refer to pulse crops, which use very little nitrogenous fertilisers.

Source: Calculated from Cassman et al. (2002).

larger variations than efficiency of crop production, and, as a result, (3) efficiency ranges of C3 and C4 biofuel chains show much larger overlap than efficiency ranges of primary crop production. This indicates that potential differences in crop production may at least partly be compensated by efficient chain organisation.

3.4 Energy use

Input use not only has implications for the impact of a given production system on its environment, but inputs that are used also represent a certain amount of energy that is invested in the production of biomass, its transport to a biofuel plant, or its conversion into biofuels. The energy may be direct – that is, use of diesel or electricity. It also involves indirect energy use, or the amount of energy that has to be invested in order to produce or distribute the input to the farmer. In practice, the amount of indirect energy can exceed direct energy use. This is especially the case in inputs that are made in energy-intensive production processes. Examples of inputs with high indirect energy requirements include nitrogenous fertilisers, agrochemicals, and machinery.

An example of energy input in a biofuel production chain is given in Table 3.3. Typical input use for soybean cultivation in the USA includes some 50 litres of diesel and gasoline per ha, plus small amounts of fertilisers (nitrogen, phosphorus, potassium), nearly 500 kg of lime, 70 kg of seeds,

Table 3.3 Typical energy use in soybean cultivation in the USA

	Unit	*Material Input*[1]	*Energy Value (MJ/ha)*
Diesel, gasoline	litre/ha	46.1	1,933
Liquid gas	litre/ha	2.0	53
Natural gas	m³/ha	4.1	161
Nitrogen	kg/ha	3.3	168
Phosphorus	kg/ha	12.1	111
Potassium	kg/ha	22.4	133
Lime	kg/ha	463.7	58
Seeds	kg/ha	68.9	324
Herbicides	kg a.i./ha	1.6	508
Insecticides	kg a.i./ha	0.04	13
Electricity	kwh/ha	17.1	127
Total			3,591

a.i. = active ingredient
[1]Calculated for the USA as an average of 19 soybean-growing states in 2006.
Source: Pradhan et al. (2011).

and 1.6 kg of agro-chemicals (active ingredients). Additional electricity use is limited to some 17 kwh per ha. The energy value of these inputs is quite variable, with most energy use being for fuels (54%), agro-chemicals (14%), and seeds (9%).

Large differences exist in the energy requirements of biofuel feedstocks. Examples of feedstocks with low energy impacts are waste materials and wheat straw or corn stover. Feedstocks with a large energy footprint include main food crops. Large differences also exist in energy use on different soil types – heavy clay soils obviously requiring more power for land preparation than sandy soils.

Furthermore, there is a crucial distinction in the energy needs of biofuel types. Biomass-to-biofuel conversion of ethanol involves steps that are energy intensive. This holds true particularly for the heating of fermented wort in order to separate alcohol from the water solution. Biodiesel production does not include this process. In some production chains, energy requirements for conversion steps are (partly) covered by the use of leftover products or waste. Cane ethanol production units in Brazil, for example, use pressed stalks (bagasse) to provide the necessary heat for the distillation process. This is done so effectively that excess electricity can be sold to the grid. The synchronous production of ethanol and electricity is referred to as *cogeneration*.

3.5 Yield levels

The evaluation of biofuel production systems requires a basic understanding of crop yields and the way they are realised. In agronomic literature (e.g., Lobell et al., 2009), different yield levels are identified. *Potential yield* is referred to as the maximum crop production level that can be realised under given conditions of solar radiation, prevailing temperature, day length, and plant genetic characteristics. This yield level can only be realised if management provides optimum conditions for crop growth; supplies sufficient water, nutrients, and optimal crop spacing; and offers protection from pests, weeds, and diseases. In reality, this is generally not the case – with the exception of conditioned growing environments like warehouses and, occasionally, under field conditions.

Under conditions of water-limited production, potential yields are reduced by water shortage. *Water-limited* yields are realised when nutrients and crop protection agents are fully available and no limitations other than restricted water availability are assumed. *Nutrient-limited* production assumes yield reduction only by limited nutrient availability, assuming perfect water supply and full crop protection.

Finally, *actual yield* levels refer to production levels where yields are reduced not only by limitation of water or nutrient availability but also by prevalence of pests, weeds, diseases, toxicities (e.g., pollution), and so forth.

This yield level is sometimes also referred to as economic production or economic potential yield level, because it is codetermined by agro-ecological conditions and economic factors such as crop market prices, costs for inputs as well as land rent, and opportunity costs of labour.

The potential crop growth rate of C3 and C4 crop types is different. It is about 200 kg of dry matter per ha per day for C3 crops in temperate regions and 350 kg for C4 crops in tropical regions. Final crop yield levels are always location specific, although some general yield indications usually can be given.

The difference between actual and potential production is generally referred to as the *yield gap*; Lobell et al. (2009) discuss the extent of this gap. As a rule, actual yield levels are largely dependent on the way farmers respond to variations in crop prices, input application costs, and risks caused by the uncertain character of weather conditions, pest and weed infestations, and market conditions. The ability of farmers to manage their crops under prevailing conditions depends on their (guaranteed) access to land, inputs, and commodity markets, and large differences exist with respect to the proportion of potential yields that are realised in the field.

An example of yield gaps is given in Figure 3.1, which depicts differences between water-limited and actual yields of winter wheat in the EU. Yield gaps are relatively small in the northwestern area of the continent as well as in parts of the UK, Spain, Italy, Poland, and Greece. The highest gaps were observed in parts of the Baltic states, Portugal, Finland, France, and Poland.

An overview of factors determining yield gaps in corn is presented in Figure 3.2. The largest gaps are found in former Eastern Bloc countries, as well as in semiarid regions of Africa and the Middle East. The smallest difference between potential and actual yields is found in western Europe. Lack of sufficient available water is the main single yield-depressing factor. It is stressed here, however, that yield gaps are not only caused by suboptimal input application levels, but that in some cases, lack of adequate knowledge or access to mechanisation is of almost equal importance. This seems to be the case in central and eastern Europe, North America, and – to a lesser extent – south and southeast Asia. Water and nutrient availability are dominant factors in Latin America and northeast Asia (including China).

Gaps between potential and actual yields tend to be smaller on good quality soils under stable weather and economic conditions, provided that farmers have guaranteed access to input and output markets (Van Ittersum and Rabbinge, 1997). Improving soil conditions or water availability can be effective to enable considerable yield improvement, as has been demonstrated on a large scale during the Green Revolution. Reducing risk levels for farmers (e.g., high output prices – or reduction of input costs) will allow farmers to invest more in crop production. This also applies to effective markets, lack of corruption, or stable economic policies. All of these factors may help to create an environment in which farmers are able to realise more of their production potential.

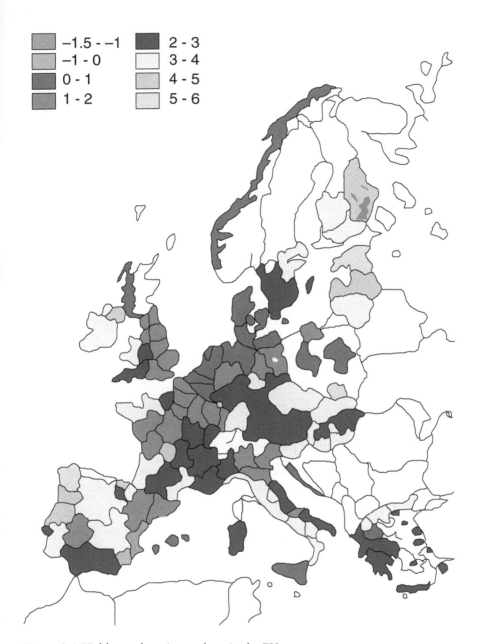

Figure 3.1 Yield gaps for winter wheat in the EU

Source: Boogaard et al. (2013).

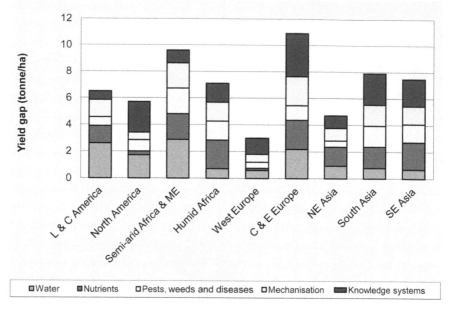

Figure 3.2 Factors explaining corn yield gaps around the world
Source: Hengsdijk and Langeveld (2009).

3.6 Quantifying biomass production

Biomass output results from a combination of factors. Several important concepts will be discussed here. Biomass production is defined by the production of area and (actual) crop yield:

$$\text{biomass output} = \text{harvested area} \times \text{crop yield} \qquad (3.1)$$

Increases in biomass output can be realised by one or a combination of four processes:

1. Expansion of arable crop area
2. Increasing harvesting frequency (Multiple Cropping Index, MCI)
3. Yield improvement
4. Replacement of crops by more productive crops

Area expansion (point 1) is considered the most damaging impact of bio-fuel production, because it may add to deforestation and the conversion of valuable vegetations. In the previous chapter it was emphasised that there is a crucial distinction between land cover and land use, whereby increases in harvested area do not necessarily lead to similar increases in land cover. This is related to changes in harvesting frequency (point 2), which were also introduced in Chapter 2. The impact of relatively small increases in MCI may be considerable.

Yield improvement (point 3) is the enhanced output of harvestable crop parts. It is generally expressed as a percentage of the preceding yield. This makes it dependent on the preceding yield level. To allow the maximum comparability of yield improvement data, we will, therefore, express them in terms of kg per ha per year.

Crop replacement (point 4) has not received much attention in biofuel policy assessment studies. This is a pity, because its impact can be considerable. In later chapters, we will see how changes in the proportions of individual crops (e.g., replacement of cotton, wheat, and barley by corn in the USA, or rapeseed pushing out cereals in the EU) can affect total biomass availability in a positive or negative way.

The combined effects of changes in harvesting frequency, yield improvement, and crop replacement on biomass availability are brought together in a new concept. We define biomass productivity (BP) as the ratio of harvested biomass to total arable area in a given region or country:

BP = harvested biomass / arable area (3.2)

BP is expressed as tonnes of biomass per ha. It usually varies between 5 and 15, with the lowest values being realised in poor or extensively managed developing countries. China is one of the countries with the highest BP, following from intensive use of agricultural inputs and frequent utilisation of double- and triple-cropping practices in its tropical rice production systems. BP is translated into biomass production by rewriting Equation 3.2:

harvested biomass = BP × arable area (3.3)

From Equation 3.3 we learn that the impact of BP can be very high when the amount of arable area to which it applies is sufficiently large.

3.7 Conclusion

This chapter introduced concepts of crop production and input use. Efficiency of crop growth is the end outcome of a number of processes. Photosynthetic efficiency is determined by light interception, temperature, and a number of crop characteristics. Considerable differences exist between C3 and C4 photosynthetic pathways. Input-use efficiency is defined for the use of nitrogen and phosphorus fertilisers, water, and agro-chemicals. Efficiency depends on the crop species, soil, and climate type as well as on the crop management system. Input use also affects the amount of energy that is needed to generate a certain amount of biomass or biofuels. High-energy needs are associated with nitrogen fertiliser applications and distillation during ethanol production. While C4 crops generally use nutrient and water inputs more effectively, differences can partly be compensated by efficient C3 biomass conversion to biofuels.

Also in this chapter, the differences between potential and actual yield levels were introduced, and it was explained how farmers can affect yield generation. This is associated not only with enhanced use of inputs but also by availability of adequate knowledge and mechanisation. As a rule, farmers tend to realise yields closer to the potential yields on good quality soils under conditions of stable weather and favourable economic conditions. Policies related to biofuel or, more generally, agricultural crop production may thus play a role in the way farmers are able to enhance biomass production levels.

Finally, this chapter explained how crop biomass production can be enhanced by expansion of agricultural land area and also by improving harvesting intensity, increasing crop yield levels, and/or replacing less productive crops with higher-yielding alternatives.

References

Bonan, G. (2002). *Ecological climatology: Concepts and applications*. Cambridge, UK: Cambridge University Press.

Boogaard, H., Wolf, J., Supit, I., Niemeyer, S., and van Ittersum, M. (2013). A regional implementation of WOFOST for calculating yield gaps of autumn-sown wheat across the European Union. *Field Crops Research*, Vol 143, pp130–142.

Cassman, K. G., Dobermann, A., and Waters, D. T. (2002). Agroecosystems, nitrogen use efficiency and nitrogen management. *Ambio*, Vol 31, pp132–140.

C_4 carbon fixation. (2013). *Wikipedia*. Retrieved 27 February 2013 from http://en.wikipedia.org/wiki/C4_carbon_fixation

Fingerman, K. R., Berndes, G., Orr, S., Richter, B. D., and Vugteveen, P. (2011). Impact assessment at the bioenergy-water nexus. *Biofuels, Bioproducts and Biorefining*, Vol 5, pp375–386.

Hengsdijk, H., and Langeveld, J.W.A. (2009). *Yield trends and yield gaps of major crops in the world*. Wageningen, the Netherlands: Wettelijke Onderzoeks Taken Natuur en Milieu.

Langeveld, J.W.A., Quist-Wessel, P.M.F., Dimitriou, J., Aronsson, P., Baum, Chr., Schulz, U., Bolte, A., Baum, S., Köhn, J., Weih, M., Gruss, H., Leinweber, P., Lamersdorf, N., Schmidt-Walter, P., and Berndes, G. (2012). Assessing environmental impacts of short rotation coppice (SRC) expansion: Model definition and preliminary results. *Bioenergy Research*, Vol 5, pp621–635.

Langeveld, J.W.A., and van de Ven, G.W.J. (2010). Principles of plant production. In H. Langeveld, J. Sanders, and M. Meeusen (Eds.), *The biobased economy: Biofuels, materials and chemicals in the post-oil era* (pp49–66). London, UK: Earthscan.

Lobell, D.B., Cassman, K.G. and Field, Chr. B. (2009). Crop Yield Gaps: Their Importance, Magnitudes, and Causes. Annual Review of Environment and Resources, Vol. 34: pp179–204.

Mekonnen, M. M., and Hoekstra, A. Y. (2010). The green, blue and grey water footprint of crops and derived crop products. In *Value of water research report series*. Delft, the Netherlands: UNESCO-IHE.

Mengel, K., and Kirkby, E. A. (1987). *Principles of plant nutrition*. Bern, Switzerland: International Potash Institute.

Norman, M.J.T., Peason, C.J., and Searle, P.G.E. (1995). *The ecology of tropical food crops*. Cambridge, Cambridge University Press

Pradhan, A., Shrestha, D. S., McAloon, A., Yee, W., Haas, M., and Duffield, J. A. (2011). Energy life-cycle assessment of soybean biodiesel revisited. *Transactions of the ASABE,* Vol 54, pp1031–1039.

Schrader, L.E. 1984. Functions and transformation of nitrogen in higher plants. In: R.D. Hauck (Ed.), Nitrogen in Crop Production. pp. 55–60.

Van Ittersum, M.K., and Rabbinge, R. (1997). Concepts in production ecology for analysis and quantification of agricultural input-output combinations. *Field Crops Research,* Vol 52, pp197–208.

Yeh, S., Berndes, G., Mishra, G. S., Wani, S. P., Elia Neto, A., Suh, S., Karlberg, L., Heinke, J., and Garg, K. K. (2011). Evaluation of water use for bioenergy at different scales. *Biofuels, Bioproducts and Biorefining*, Vol 5, pp361–374.

4 Assessing greenhouse gas emissions

J.W.A. Langeveld

4.1 Introduction

Few elements of the production of biofuels have drawn so much attention and caused so much debate as their environmental performance. Predominant in the assessment of biofuels' performance is their contribution to the reduction of greenhouse gases (GHG). This is not surprising, as the introduction of biofuel-supporting policies in many cases was underpinned by their role in lowering CO_2 emissions (i.e., by replacing fossil fuels). It is noted that many years after biofuel policies were introduced in the USA, EU, and many other industrial and emerging countries, there still is much confusion on how to assess the impact of biofuel production chains on GHG.

One of the most important explanations for why there is so little agreement on methodology is the complex and integrated character of biofuel production chains. Assessing GHG emission reductions requires an examination of the full life-cycle emissions of the fuel, thus evaluating land-use changes, land management practices, biomass feedstock production and its transportation to the plant, and biomass conversion processes. Basically, three steps need to be considered: (1) land preparation, (2) crop cultivation, and (3) biomass conversion and biofuel application. Each of these steps involves different technologies, disciplines, actors, and stakeholders. They require different analytical approaches.

Land preparation refers to the selection and preparation of a piece of land where the biofuel feedstock is to be cultivated. Biofuels in theory can be produced from crop residues, industrial residues, forest residues, or household or industrial waste materials. In practice, however, most biofuel production still involves primary food crop biomass. This means that agricultural plots need to be selected and cultivated.

Basic concepts related to land cover and land use were introduced in Chapter 2, where it was explained that the more dramatic impacts of biofuel expansion are related to changes in *land cover* (which is related but not equal to *land use*). Issues of deforestation and the conversion of other nature areas or grasslands all have serious consequences for the living conditions of the local population and the conservation of biodiversity as well as of carbon stocks. Changes in land cover can be monitored by using satellite images. Land conversion tends to be an absolute process. The impact on GHG emissions is largely defined by changes in available carbon stocks.

Biofuel expansion can, however, also be related to changes in cultivation intensity which usually have a more gradual character. They normally are observed by comparing cropping calendars and other elements of land use over a range of years. Evaluating changes in cultivation intensity requires availability of land-use data in time series. Such data are rather scarce, or at least less accessible for researchers or analysts, while quality often is problematic.

Cultivation intensity is only one of the elements determining biofuel crop cultivation. Other elements refer to input use and the way soils, water, and air are affected. Crop cultivation practices tend to show large variation because they depend on the way farmers respond to (variations) in local weather, soil, plant health, and economic conditions. Cultivation is studied by agronomists, whose reports and papers often have limited accessibility for non-specialists – as is the case with most scientific work – especially those working from abroad. Most focus on technical issues related to single crops, crop varieties, soils, or cultivation activities.

Because there are so few generic agronomic publications, it is hard for non-agronomists (e.g., life-cycle assessment [LCA] specialists) to get a well-balanced overview of crop cultivation practices in a given area and the way this is affected by (changes in) local and national biophysical, economic, and social conditions. An exception to the rule is the book on bioenergy crops by El Bassam (2010).

The final source of variation in GHG balance calculations is based in the process of biomass-to-biofuel conversion and in biofuel application. Conversion technologies get remarkably little attention in the biofuel debate, which tends to overlook both historical improvements that have been realised in biofuel production (e.g., enhanced efficiencies in cane-to-ethanol conversion that were developed in Brazil) and future developments that have been projected (for an overview of the latter, see Burrell, 2010). With some exceptions, biofuel literature does not refer to the combustion process (i.e., the way biofuels are burned in car engines). Instead, analyses usually look no further than producing the feedstock, converting it into biofuels, and distributing this to cars. Most LCA studies use the car tank as system boundaries.

Biofuel production chains have many options to adapt to changes in prevailing economic and biophysical conditions. This book does not aim to list them all. Nor does it intend to discuss all possible variations that can affect GHG emissions of biofuel production and utilization. To understand the debate on GHG performance – and the large variations therein – it is, however, important to be aware of the way specific activities in the biofuel production chain affect its carbon impact.

The current chapter discusses ways biofuel producers (involved in the production of biofuel feedstocks and their conversion to fuels) can respond to changes in related policies and market conditions. Variations in biofuel production are discussed in Section 4.2. Principles of GHG calculations are presented in Section 4.3, and common calculation tools and models used in emission calculations are covered in Sections 4.4 and 4.5, respectively. Section 4.6 discusses issues related to the modelling of land-use change.

4.2 Variations in biofuel production

Biofuel production chains are complex and involve many stakeholders and actions; chain organisation tends to be complex. Consider, for example, different steps in the production process presented by Dale et al. (2013). Five basic steps are taken (Figure 4.1). Feedstock production refers to plot selection and crop cultivation. Feedstock logistics involve harvesting, collecting, and processing of the feedstocks, including storage and transport to the conversion plant. Conversion involves the actual production of the biofuel and possible co-products. Biofuels then are distributed, which usually requires transport and storage. Final use includes blending of the biofuel with fossil counterparts.

Each of these steps may involve different stakeholders and production locations, offering options for input use and technological applications. The large number of options that usually are available may provide necessary flexibility and room for adaptation to prevailing (economic) conditions. Some of these adaptation processes have been described by Langeveld and Sanders (2010). Following the introduction of biofuel policies, a chain of reactions may follow.

Expanding biofuel production requires extended production or import of crop feedstocks. This leads to an increased demand for land and inputs and consequently affects feedstock availability at both the local and international levels. As biomass markets around the world are linked, this will, at

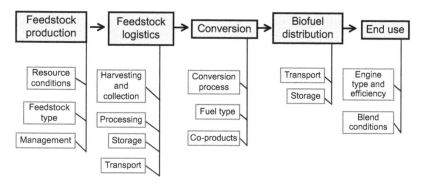

Figure 4.1 Phases and activities in the biofuel production process
Source: Dale et al. (2013).

some point, change conditions for traditional biomass users, including food and feed production, as well as that of fibres, pulp, and paper. The demand for land will increase, which makes it likely that land use will become more intensive (reduce fallow, increase multiple cropping). Also, expansion of arable area may be expected.

Many of the effects, in their turn, will have implications of their own. Increased demand for biofuel crops and enhanced availability of co-products, for example, will affect animal feed markets. Such indirect effects can, however, only be identified by specialists (and even for them they generally are difficult to link to a specific cause). Thus, while a complete assessment of biofuel expansion requires an evaluation of all its effects, be it direct (and clearly visible) or indirect (much less visible), nobody seems to be able to grasp all its implications. This makes the debate on expanding biofuel production and its impact (on food prices, land-use changes) very complicated.

Biofuel production will, thus, affect several aspects of day-to-day life. As different elements of society are influenced, society as a whole will have to respond. This may lead to changes in related policies, and can go as far as imposing restrictions on biofuel production or bans on food exports, as recent examples have shown. An overview of changes in biofuel production, market conditions, consumer behaviour, and policy that were provoked by the introduction of biofuel policies is given in Table 4.1.

Many of the processes described in Table 4.1 indeed have been observed. Demand for biofuel crops has increased, as has availability of co-products. Input use and multiple cropping have been raised; land requirements and crop yields have gone up. Changes in commodity prices have caught the attention of researchers, NGOs, the public, and, consequently, policy makers. As a result of this, biofuel policies have been adapted. As we have seen, the way in which they were formulated differs greatly among countries. Blending obligations, regulations on feedstocks that can be used, price instruments, subsidies, import levies, and export subsidies tend to be selected by governments according to prevailing conditions in local energy, fossil fuel, and agricultural, food, and land markets. Moreover, they tend to be adjusted following changes in food and biofuel production, fossil prices, and food prices.

Public debate tends to play a significant role in both formulation and adjustment of biofuel policies. Two factors play a dominant role. The first is the contribution of biofuels in reducing GHG emissions. The second is the impact of (emerging) biofuel production on the use of land and other resources and how this affects food availability. The way these factors affect biofuel policy making is best demonstrated by the way biofuel policies are adjusted in the USA, EU, and China.

After the Searchinger paper indicated how carbon emissions could be affected by induced land-use change, it became one of the most contentious issues (along with the food vs. fuel debate) relating to biofuels debated in the media, scientific journals, and magazines (Searchinger et al., 2008). This

Table 4.1 First- and higher-order effects of increased biofuel production

	Farmers and Biofuel Producers	Markets	Consumers	Public Opinion and Policy
1st order: Increased biofuel demand	1. Increased demand for food/feed crops			
2nd order: Market response		2. Reduced crop surpluses 3. Crop price increases		
3rd order: Consumer response			4. Worries about food price increases	
4th order: Indirect market effects, policy reactions	5. Crop area expansion, increased input use. Reduce fallow, increase multiple cropping	6. Price increases for land and inputs, reduce prices of co-products		7. Call for regulation of crop use, land use
5th and higher order: Reactions to lower-order effects	8. Improve input use efficiency; select more productive crops; use cheaper feedstocks or more efficient conversion techniques; reduce co-products	9. Increase use of co-products; replace food crops with nonfood crops or crop residues	10. Change consumption patterns (cheaper food, less meat)	11. Debate on ethics of biofuels made from food crops; change policies

Source: Adapted from Langeveld and Sanders (2010).

controversy intensified in 2009 when the California Air Resources Board (CARB) set rules that included ILUC impacts to establish the California Low-Carbon Fuel Standard that entered into force in 2011 ('Indirect Land Use Change Impacts of Biofuels', 2012). In May of that year, the US Environmental Protection Agency (EPA) released a notice of proposed rulemaking for implementation of the 2007 modification of the Renewable Fuel Standard (RFS). The EPA's proposed regulations also included ILUC, causing additional controversy among ethanol producers. The EPA's February 2010 rule incorporated ILUC based on modelling that was significantly improved over the initial estimates ('Indirect Land Use Change Impacts of Biofuels', 2012).

The UK Renewable Transport Fuel Obligation program requires the Renewable Fuels Agency (RFA) to report potential indirect impacts of biofuel production, including indirect land-use change or changes to food and other commodity prices. A July 2008 RFA study found several risks and uncertainties, concluding that further examination to incorporate indirect effects into calculation methodologies was needed. A similarly cautious approach was followed by the EU. Following the introduction of biofuels policies earlier that year (2008), in December the European Parliament adopted more stringent sustainability criteria for biofuels and directed the European Commission to develop a methodology to factor in GHG emissions from indirect land-use change ('Indirect Land Use Change Impacts of Biofuels', 2012). That methodology finally was proposed in 2012.

In China, where the GHG impact has never been part of the justification for biofuel development, preliminary biofuel policies were adjusted when the government concluded that initial biofuel production claimed too much of the available corn and soybean stocks.

4.3 Quantifying emissions

An example of a GHG balance is given in this section. Under the Renewable Energy Directive (RED), GHG emissions from the production and use of biofuels are calculated as follows:

$$E = e_{ec} + e_l + e_p + e_{td} + e_u - e_{sca} - e_{ccs} - e_{ccr} - e_{ee} \qquad (4.1)$$

where:

E = total emissions from the use of the fuel
e_{ec} = emissions from the extraction or cultivation of raw materials
e_l = annualised emissions from carbon stock changes caused by land-use change
e_p = emissions from processing
e_{td} = emissions from transport and distribution
e_u = emissions from the fuel in use

e_{sca} = emission saving from soil carbon accumulation via improved agricultural management
e_{ccs} = emission saving from carbon capture and geological storage
e_{ccr} = emission saving from carbon capture and replacement
e_{ee} = emission saving from excess electricity from cogeneration

Emissions from the manufacture of machinery and equipment are not taken into account.

GHG savings now are calculated as follows:

$$(EF - EB)/EF \qquad\qquad (4.2)$$

where:

EB = total emissions from the biofuel
EF = total emissions from the fossil fuel that is replaced

The total emission and the way it is composed of the different elements strongly depends on the type of feedstock; the way it is cultivated, transported, and converted; and the distribution and final use of the biofuel, as well as its (indirect) impacts. Steps that contribute the most to the final emission level include plot preparation (e.g., removal of previous vegetation), the use of nitrogenous fertiliser during crop cultivation, energy use during conversion (for ethanol production), and impacts of land-use change.

In practice, most biofuels are still produced from primary crops that are cultivated on existing arable plots. An example of such a situation is given in Table 4.2, which presents GHG emissions during the production of sugar beet ethanol in the Netherlands. Most emissions are caused during beet-to-ethanol conversion. For this study, one conversion model has

Table 4.2 GHG emissions from beet ethanol production in the Netherlands
  ($g\ CO_2$-eq./MJ)

	Crop Cultivation	Field to Plant	Conversion	Distribution	All
Clay soil: 2005	6.3	2.7	33.5	0.3	42.7
Clay soil: 2006	8.2	2.6	33.5	0.3	44.5
Clay soil: 2007	11.2	2.7	33.5	0.3	47.7
Average	8.6	2.7	33.5	0.3	45.0
Sandy soil: 2005	4.5	2.7	33.5	0.3	40.8
Sandy soil: 2006	3.6	2.6	33.5	0.3	40.0
Sandy soil: 2007	3.6	2.5	33.5	0.3	39.8
Average	3.9	2.6	33.5	0.3	40.2

Source: De Visser et al. (2008).

been used – hence conversion efficiency is equal for all cases. Two-thirds of emissions are associated with biomass conversion. Transportation from the field to the factory and distribution of the ethanol cause only very limited emissions.

Data from Table 4.2 show that beet production on clay soils requires more energy than production on sandy soils because diesel requirements for land preparation, and so forth, are higher. Also, beets on sandy soils received more manure, which reduced the need for nitrogen fertilisers. Furthermore, in 2007 yields on clay soils were low and inputs applied had to be divided on fewer MJ of ethanol.

4.4 Emission modelling

Over time, a considerable number of GHG calculation tools and models have been developed. Many were developed in scientific institutions. Some of them are discussed in this section.

The Global Emission Model for Integrated Systems (GEMIS) is a life-cycle analysis program and database that was developed at Öko-Institut and Gesamthochschule Kassel. The database offers information on fuels, processes, materials, transport, and so on. GEMIS provides a full-life assessment but can also be used to analyse costs of fuels and energy systems.

GHGenius was developed for Natural Resources Canada, based on the Lifecycle Emissions Model (LEM) by Mark Delucchi. It analyses emissions associated with the production and use of transportation fuels, estimating life-cycle emissions of primary GHG and pollutants.

Greenhouse gases, regulated emissions, and energy use in transportation (GREET) models are useful in calculating the petroleum and GHG footprint of transport or power generation; comparing different alternative fuels and vehicle technologies for a future medium-duty, heavy-duty, or off-road vehicle purchase; comparing the automobile's footprint to other transportation modes; viewing alternatives for reducing the impact on the environment; and so on.

GREET is a publicly available LCA tool for consistently examining life-cycle energy and emissions of vehicle and fuel systems that was developed by Argonne National Laboratory in the USA. It covers more than 85 vehicle and fuel systems, major fuel types, and vehicle technologies; it generates assessments of major GHG emissions (CO_2, CH_4, and N_2O), air pollutants (VOC, CO, NOx, SOx, particulate matter), and energy use (total energy, fossil energy, petroleum energy).

Following RED methodology, the European BioGrace project developed an Excel tool that can depict emissions related to biofuel production. The complete BioGrace GHG emission calculation tool includes 22 biofuel production pathways given in the RED, a calculation sheet on annualised emissions from carbon stock changes caused by land-use change, a calculation sheet on emission saving from improved agricultural management, a sheet for the calculation of N_2O field emissions, and an overview of calculation rules. Currently, a special tool is being developed to calculate GHG emissions from solid biofuels.

Footprint Reporter is an easy carbon footprinting tool that can help organisations measure, report, and reduce their emissions. This off-the-shelf system has been developed to help small- and medium-sized businesses measure their carbon footprints and comply with government carbon-reporting standards. It also provides options for benchmarking to improve efficiency and cut costs. Customised versions are available and are provided by Best Foot Forward consultants (http://footprintreporter.com).

One recent tool developed on behalf of Unilever (the Cool Farm Tool) focuses on GHG emission calculations for farming. This tool gives instant results and invites users to look for alternatives with lower emission levels. The tool, developed by the University of Aberdeen, is aimed at farmers, supply chain managers, and companies interested in quantifying their agricultural carbon footprint. It calculates the GHG balance of farming, including emissions from fields, inputs, and land-use change using Intergovernmental Panel on Climate Change (IPCC) Tier 2–type methodology.

GHG emissions are also calculated using a number of existing economic models run by major economic research centres. These models are based either on partial equilibrium (PE) or computable general equilibrium (CGE) models. Their aim is to evaluate policy options using well-known principles of price relationships between crop supply and demand. PE models assume that prices and quantities in other markets remain constant, which makes them better suited to assess short-term responses to shocks. PE models provide a more detailed description of a given sector of interest but do not account for the impact on other sectors. Several examples of PE models used in the assessment of the impact of biofuel development include AGLINK/COSIMO (COmmodity SImulation MOdel), ESIM (European SImulation Model), FAPRI (Food and Agriculture Policy Research Institute), and the IMPACT (International Model for Policy Analysis of Agricultural Commodities and Trade) model (Al-Riffai et al., 2010).

CGE models simultaneously consider linkages between all sectors in the economy. The modelling framework provides an understanding of the impact of biofuels on the overall economy by accounting for all the feedback mechanisms between biofuels and other markets, and by capturing factor market impact. This allows CGE models to provide a better global assessment of policy or market changes, taking linkages in the economy into account and predicting outcomes that are more representative of medium- and long-term impact (Al-Riffai et al., 2010).

Based on the way in which bioenergy is treated, three types of CGE models can be identified (Al-Riffai et al., 2010). The *implicit approach* employs an ad-hoc procedure of determining the quantities of biomass necessary to achieve certain production targets. An example of this approach is given by Banse et el. (2010) in their extended version of the Global Trade Analysis

Project (GTAP-E) CGE model. Ethanol is substituting vegetable oil, fossil oil, and petroleum products. It is produced only from crop inputs (sugarcane\beet and cereals), thereby capturing only a part of ethanol production technology.

The *latent technology approach* focusses on existing production technologies that are not implemented as active during the base year but can become active at a later stage. Information on the inputs and cost structures of the different types of biofuels are required in modelling latent technologies. The third category of CGE biofuel studies include studies that *disaggregate bioenergy production*. Al-Riffai et al. (2010) introduced biofuels in the GTAP database, which provides the underlying structure to global CGE models. Since bioenergy sectors are not explicitly identified in this database, biofuel production is introduced in a generic way. Production of ethanol (from coarse grains and sugarcane) and biodiesel (from the oilseeds sector) from existing food sectors in the GTAP 6 database are determined using external data on biofuel production, cost, and trade.

Khatiwada et al. (2012) compared outcomes of GHG emissions of Brazilian cane ethanol calculated by main American and European models and calculation tools. They seem to come to more or less similar results, emissions averaging around 24 g CO_2-eq./MJ (Table 4.3). Emissions calculated under the USA-EPA model are higher, especially emissions related to crop cultivation. Generally, crop cultivation is responsible for 60% (European tools) to 100% (USA tools) of total net emissions. In some models, conversion causes a net reduction of the emissions because of fossil fuels that are replaced by electricity that is produced (co-generation) from bagasse (cane fibres) that burnt during the conversion process.

In practice, variation between different GHG emission calculations tends to be much higher than the results from Table 4.3 suggest. It is not uncommon for some of the studies to suggest that considerable emission reductions are possible while others conclude that emissions in fact exceed those of fossil fuels. Emissions related to land-use change (ILUC) play a dominant

Table 4.3 GHG emissions from ethanol production in Brazil (g CO_2-eq./MJ)

	USA-EPA	CA-CARB	EU-RED	UK-RTFO
Crop cultivation	36.3	18.6	14.1	14.5
Field-to-plant transport	−1	−1	0.8	0.9
Conversion	−10.5	−5.1	0.9	0.6
Ethanol distribution	4.1	6.1	8.1	8.0
Biofuel end use	0.8	0.8	−[1]	−[1]
Total	31.0	20.4	24.0	24.0

[1]Included in 'ethanol distribution'.

Source: Khatiwada et al. (2012).

role in this. Estimates for Brazilian cane ethanol presented by the tools listed previously vary between 0 g CO_2-eq./MJ for UK-RTFO and 46 g for CA-CARB (Khatiwada et al., 2012). This makes land-use change in most cases the largest individual emission factor.

4.5 Modelling land-use change

Modelling land-use change apparently is more complex than modelling biofuel production. Croezen et al. (2010) compared different approaches assessing ILUC and its impact on GHG emissions. They identify two main approaches: (1) the use of economic models and (2) the use of generic ILUC factors to be applied in GHG balance calculations. The authors compare outcomes of major biofuel modelling exercises (including work by the International Institute for Applied Systems Anaysis [IIASA], the International Food Policy Research Institute [IFPRI], the Joint Research Centre of the European Commission [JRC], Purdue University, and the Agricultural Economics Research Institute [LEI]), presenting ranges of calculated GHG emissions caused by ILUC.

More details of the models are provided in a JRC study (Edwards et al., 2010). Model outcomes are compared by expressing land-use increases due to biofuel expansion as marginal values (extra land needed for a given unit of biofuel energy), thus ruling out differences in assumptions, time frame, and so on. An overview of the outcomes is presented in Table 4.4.

LEITAP generally provides the highest land requirement assessments and impacts the lowest. Models do not only show large differences in estimated land requirements, but they also differ as to which region is to provide necessary crop feedstocks. Compare, for example, the share of land needed for the production of EU ethanol in the USA and the EU for GTAP (794,000 vs. 165,000 ha) to those given by other models (which more evenly distribute land use in the EU and the USA, thus requiring more imports for the EU).

Table 4.4 Extra land needed to generate 1 million tonne of oil-equivalent (MToe) of energy to fulfill EU biofuel targets (thousand ha)

	Ethanol		*Biodiesel*			*Model Owner/Operator*
	EU	*USA*	*EU*	*USA*	*Far East*	
Aglink	574	510	230	242		OECD, France
FAPRI	394		435			CARD, USA
Impact	116–223	107–223				IFPRI, USA
GTAP	794	165	376		82	Purdue, USA
LEITAP	731	863	1928		425	LEI, the Netherlands

Source: Calculated from Edwards et al. (2010).

Burrell (2010) provides a detailed comparison of three models: Aglink, ESIM (Banse), and CAPRI (Common Agricultural Policy Regionalised Impact). The models were used to assess the impacts of the current EU bio-fuels policy on land use and on agricultural production. Results and model specifications are compared. They show large differences with respect to level of aggregation (commodities, land units), which makes it difficult to compare spatial outcomes in a disaggregated way. There are also different methods to incorporate policies, or biofuel supply and demand, or in the way by-products are generated during biofuel production.

Given these differences, it comes as no surprise that modelling outcomes seem insufficiently unanimous as to predict what (indirect) land-use effects the EU policy will have. Large differences are found as to what areas will be needed – both within and outside Europe, what crops will have to be produced, and what GHG effects displacement of current food production may be expected.

4.6 Conclusion

This chapter discussed GHG calculation methods. Factors that cause varia-tions in emissions of biofuel production chains relate to land preparation, crop cultivation, and biomass conversion, while indirect effects on land use may be of predominant importance. Production chains are complex and show sufficient room to adjust to changing economic or legal conditions, and these adjustments may affect GHG emissions of the end product.

GHG emissions can be calculated using a range of tools and models, most of which originate from scientific backgrounds. Modelling emissions related to land-use change can be done by economic models or by applying so-called ILUC factors. Given the importance of emission reductions in the justifica-tion of biofuel policies, it is of crucial importance that both their impacts on land use and the effects this has on biofuel emission reductions are calcu-lated in a scientifically solid way.

Estimations for land-use requirements presented by different models sug-gest there is little debate on the total amount of land needed to realise biofuel targets set out in EU policies. Estimates of ILUC-based emissions presented for Brazilian cane ethanol, however, show these are not only large but also highly variable.

The economic models used to assess land-use requirements and GHG emissions all have been designed for other purposes, and biofuel production was introduced only after development and calibration of the original mod-els. Also, the outcome of ILUC-related emissions – in some cases suggesting that up to 100 years of biofuel production will be needed to compensate for carbon releases due to area expansion – and the role they play in public and policy debate in many countries further underpins the need for a solid analysis.

In this book, a different approach is chosen. It describes land-use dynamics and crop cultivation practices in biofuel-producing nations in much more detail than normally is done. This information is used to assess land-use changes that have been provoked by biofuel production. The analysis is based on monitored changes in biofuel crop production of the (recent) past, which allows us to evaluate the outcome in light of other changes in land use that have been observed.

References

Al-Riffai, P., Dimaranan, B., and Laborde, D. (2010). *Environmental impact study of the EU biofuels mandate.* Washington, DC: International Food Policy Research Institute.

Burrell, A. (2010). Impacts of the EU biofuel target agricultural markets and land use. A comparative modelling assessment. Seville, Spain: Institute for Prospective Technological Studies (ITPS).

Croezen, H.J., Bergsma, G.C., Otten, M.B.J., and van Valkengoed M.P.J. (2010). *Biofuels: Indirect land use change and climate impact.* Delft, the Netherlands: CE Delft.

Dale, V.H., Kline, K.L., Perla, D., and Lucier, A. (2013). Communicating about bioenergy sustainability. *Environmental Management,* Vol 51, pp. 279–290. DOI 10.1007/s00267-012-0014-4

De Visser, Chr., van de Ven, G., Langeveld, H., de Vries, S., and van den Brink, L. (2008). *Duurzaamheid van ethanolbieten* [Sustainability of ethanol beets]. Lelystad, the Netherlands: Accres.

Edwards, R., Mulligan, D., and Marelli, L. (2010). *Indirect land use change from increased biofuels demand.* Ispra, Italy: JRC.

El Bassam, N. (2010). *Handbook of bioenergy crops: A complete reference to species, development and applications.* London, UK: Earthscan.

Indirect land use change impacts of biofuels. (2012). *Wikipedia.* Retrieved 12 February 2012 from http://en.wikipedia.org/wiki/Indirect_land_use_change_impacts_of_biofuels

Khatiwada, D., Seabra, J., Silveira, S., and Walter, A. (2012). Accounting greenhouse gas emissions in the lifecycle of Brazilian sugarcane bioethanol: Methodological references in European and American regulations. *Energy Policy,* Vol 47, pp384–397.

Langeveld, J.W.A., and Sanders, J. (2010). General introduction. In H. Langeveld, J. Sanders, and M. Meeusen (Eds.), *The biobased economy: Biofuels, materials and chemicals in the post-oil era* (pp3–17). London: Earthscan.

Searchinger, T., Heimlich, R., Houghton, R.A., Dong, F., Fabiosa, J., Tokgoz, S., Hayes, D., and Hsiang, Y.T. (2008). Use of U.S. croplands for biofuels increases greenhouse gases through emissions from land-use change. *Science,* Vol 319, pp1238–1240.

5 How land rights influence the social impacts of biomass production

Th. Hilhorst

5.1 Introduction

Increasing biofuel production of biomass may result in an expansion and intensification of biofuel crop area. Arable land with the required qualities (soil quality, water availability, market access) is a finite resource. Expansion may imply conversion of land now under (protected) natural forest, wetlands, or grazing lands. Expansion can also lead to competition with existing producers and their farming and livestock systems. Competition can also emerge over water. Particularly in countries where land and water rights protection is weak, expansion of lands under biomass can cause displacement of smallholders, herders, and other groups who base their livelihoods on the resources taken over for biomass production. These challenges are relevant for all investments that imply taking over control of large tracts of land or other natural resources. The difference for biomass production is that these enterprises may be politically more sensitive when perceived as displacing food production and increasing food insecurity.

This chapter analyses how land rights affect land availability for expanding biomass production as well as the impact of biomass production on the local economy and on social relations. The focus is on developing countries, particularly those in Africa. An overview is presented of challenges regarding tenure rights and arrangement with respect to contracting, employment, and benefit sharing that influence the effect on local economic development and the social impact of investments. The chapter first discusses the relationship between land scarcity and land and resource rights (Section 5.2). Then it determines land tenure implications (Section 5.3) and the social impacts (Section 5.4) of biofuel feedstocks.

5.2 Land scarcity and land and resource rights

Land governance

In most countries the "land sector" is characterised by institutional fragmentation where governmental responsibility for land is spread over a large number of public institutions at national and sub-national levels, which are often poorly coordinated. An additional complexity is that in many countries foreign direct investment in land is handled by a special agency in charge of promoting investments without much involvement of agencies in charge of agriculture or natural resource use. As already alluded to, the gap between legal provisions, actual implementation, and local legitimacy of the regulatory framework is often enormous. This is often the case for land acquisitions in a context of customary land tenure. Weak land governance implies that the rights of local populations and investors are not protected. This situation adversely affects investment, economic growth, sustainable use of the environment, and social stability. Such dilemmas are less likely when the lands acquired are former state farms (assuming that there are no conflicts around the expropriation which may flare up when a state land is privatised) or involve established commercial farms that come up for sale. The more risk-averse investors, with a more long-term perspective, will avoid situations of high-tenure insecurity and seek properly zoned agricultural land (which may be underpriced).

Land governance is defined as the process by which decisions are made regarding the access to and use of land and natural resources, the manner in which those decisions are implemented, and the way that conflicting interests are reconciled (Palmer et al., 2009).

Key aspects of the institutional arrangements governing land are as follows: (1) the legal and institutional framework around rights recognition and enforcement, (2) definition of public land, (3) public availability of reliable land information, (4) regulations and systems for land-use planning and land management, (5) mechanisms for dispute resolution and conflict management, (6) land taxation, and (7) regulations around large-scale land acquisitions.

International controversy around large-scale land acquisition

The local and global demand for land continues to increase. As a result of demographic growth and higher commodities prices more land is needed to produce food, fuel, and fibre for establishing environmental amenities and to provide land for urban expansions, infrastructural projects, extractive industries, and so on.

A remarkable global trend is a renewed interest in acquiring large tracts of land for farming picked up in the mid-2000s, partly in response to deregulation (taxes, exporting profits) that benefitted foreign companies in particular. Other drivers were investors and investment funds looking for new (undervalued) assets (such as land in developing countries and the former USSR) and expansion of biomass production in response to regulation on biofuels in the USA and the EU. After the financial crisis in 2007 and growing concerns over global food security since 2008 – when prices for agricultural commodities in general picked up, the interest of agribusiness to move into primary production further increased. This drive has been intensified by governments in developing countries that are actively promoting (foreign) direct investment in farmland and large-scale industrial farms. Biomass production may have been a frontrunner, and jatropha plantations in particular did become popular, although many failed later on, but food production also has become important. Many of the crops involved are actually multipurpose and can be used for food and biomass (sugarcane, palm oil, soy).

Large-scale land acquisitions are, however, controversial and regularly attract media headlines, particularly when situated in food-insecure countries in Africa and Asia. The start of the media's interest was the Daewoo case in Madagascar. Daewoo is a South Korean company that announced having acquired a lease over more than 1 million ha of land in Madagascar to produce maize and other crops, which was reported in November 2008 in the *Financial Times*. The announcement provoked an outcry in Madagascar and contributed to the toppling of the regime then in power; the Daewoo deal did not go ahead but it did not stop this type of investment. Other large-scale plantations were established later in Madagascar.

Large-scale land acquisitions also have become a source of controversy at the international level. The large-scale land acquisitions triggered discussion at the United Nations (UN) and led to broad international civil society mobilisation, a flurry of research, and the formulation of some principles for Responsible Agro Investments (RAI).

The controversy also is one of the reasons for the endorsement in 2012 of the "voluntary guidelines for the responsible tenure of land, fisheries and forestry in the context of food security" by the Committee on World Food Security (CFS). The CFS continued with an analysis of responsible agro-investments, including business models that do not result in permanent loss of land rights for communities, enabling their economic participation in new enterprises and forms of benefit sharing.

There is reason for concern. Local populations become aware only when the actual farm work starts. In cases of displacement, the populations concerned may not be compensated at all, or the money is insufficient to buy new land (de Schutter, 2011; Oxfam International, 2011; HLPE, 2011). Some civil society and UN organisations' published documents indicate that even though they welcome private investments in agriculture, the transfers of large tracts of land to companies for large-scale farming, forestry, mining, and other forms of resource use are causing displacement, local food insecurity, conflict, and environmental degradation.

Therefore, the controversy over large-scale land acquisitions is understandable. Even when the transfer of land is legal, large-scale land acquisitions are perceived locally as de facto 'land grabs' when the legitimate rights of existing land users are ignored. The local population living and working on these lands, often already for generations, considers themselves as the legitimate right holders. But because of transfers, they no longer have access to forests, grazing, or water resources. Some land users are even evicted or (forcefully) resettled.

Moreover, much transferred land has not been developed at all and may have been acquired for speculation purposes; inventories from around 2010 estimated that this was the situation for roughly two-thirds of the land being transferred, but available data are weak. There is a type of investor who seeks out situations of high risk and potentially high profit, and therefore focusses either on countries where governance is weak or on post-conflict countries (Morris et al., 2009). There are also many cases of bankruptcy, but provisions for the land to return to the state or the original users may be lacking.

Where investment did actually take place, the first results of these farms – in terms of productivity and created employment – are not encouraging, although it will take some time before agribusiness companies are fully operational.

The complexity of local land rights and the myth of empty lands

Empty, vacant lands without owners do not exist. Those lands (and water) are used by smallholders or herders, who in many parts of the world are not immediately visible. The land and natural resources on the land are important assets for local livelihoods and contribute to food security. These lands also produce ecosystem services such as hydrology, soil protection, and bio-diversity. Therefore, the assumption should be that all land is in use, even when no traces of fields or fences are visible to the inexperienced eye. Fallow lands, for example, resemble forestlands: the land is not farmed for a number of years, then the vegetation is burned and in the fertile ashes crops are grown. Shifting cultivation is still common because smallholders have limited access to manure and fertilisers. Therefore, farmers use fallow lands to restore soil fertility.

Herders use pasture lands only part of the year. Grazing lands are seldom fenced and when they are, it is to protect crops against herds. Also, groups living elsewhere can have user rights and use the land for part of the year; for example, livestock holders, fishermen, users of forest produce, or hunters (see Alden Wily, 2010).

As indicated previously, competition over water for irrigation or water that is captured in the soil (e.g., green or 'virtual water') between biomass production and smallholder production systems may be an additional challenge, particularly in areas with low rainfall access to groundwater or where irrigation is essential for producing. There are cases in Mali, for example, where upstream sugarcane plantations compete with downstream smallholder irrigation schemes. In addition, the contract signed between the company and government guarantees access to sufficient water for irrigation. This competition over water may become a source of conflict (Baumgarten, 2012).

Customary land rights

In a number of countries, legal ownership of all land, farmed or not, is vested in the state, acting as trustee or steward on behalf of the population. Farmers only have user rights. There are also countries where unfarmed land and forests are claimed by the state as public lands. However, customary tenure systems dominate in rural areas, even when not officially acknowledged. The result is overlapping formal and informal land tenure systems (legal pluralism), which is a source of insecurity for local land users and companies.

Situations of legal pluralism have consequences for investors in large-scale biomass production that are important for the sustainability and profitability of the enterprise. In countries where customary land tenure systems and rights to pastures and forests are not recognised in law, investors in biomass seeking land for expansion face a complex land tenure system. If the land acquisition is not accepted locally, the investors are met with resistance.

Assurance by a national government is no guarantee for local acceptance of the land acquisition and peaceful relations with neighbouring communities. This is another challenge for investors. Government-acquired land offered to companies may be obtained either by invoking its legal right over the land (crown land, trustee land, public land, etc.) or by actual expropriation. The government may even have played a key role in negotiations over community land. Although the transaction is legal, there is no guarantee that the local population agreed or even that apparent acceptance is the result of free and prior informed consent. Therefore, relying only on government information and procedures can be risky. Companies must use due diligence when looking into how the land was acquired and researching community perception and benefits.

Of course, when companies are participating actively in the acquisition of the land, they are directly responsible for the quality of community consultation as well as other preparations such as social and environmental impact assessment. A process that respects international standards for responsible investment is increasingly required by lenders and in international supply chain management. If the process of acquisition and contracting does not meet the standards of responsible investments, such as those set by the International Finance Cooperation (IFC), commodity round tables, and so on, the company cannot secure finances from these funds and may have to deal with reputational ruin if land conflicts erupt.

Business models involving smallholders

Land tenure-related risks for investors in biomass will be reduced when using business models that rely on smallholders or leasing their land. Smallholders will then produce biomass crops and sell the harvest to processors. The relation with smallholders can be intensified by working with contracts and outgrower schemes and moving from supply changes to value changes.

Socioeconomic effects depend on the profitability of the crops, risks, employment, and how the crop fits in the production system. The effects on social and economic developments depend also on the type and number of farmers involved and whether tenure relations change.

Leasing land from smallholders may also reduce the negative impact of large-scale farming, depending of course on the conditions of the contract. There is no benefit for smallholders who leased all their land to a sugarcane plantation, have no more land left for farming, and cannot live off the rent paid by the plantation, while the employment generated is limited in its number of days and is poorly paid.

Business models involving land for biomass production in areas where there is limited land available to companies are the only way for expanding production (e.g., palm oil expansion in west Africa).

5.3 Land tenure implication of various biomass feedstocks

Understanding land tenure and water rights implications of crops is particularly important for investors in perennial crops that will occupy the land for decades, or when significant investments take place in soil preparation and infrastructure, such as irrigation or roads. An understanding of tenure systems is equally important for companies that invest in the processing of perennial biomass, such as sugarcane, that is produced by smallholders.

Examples of perennial crops are oil palm, Jatropha, and trees grown for pulp. Others are technically not perennial crops but occupy land for several decennia (e.g., sugarcane). Biomass products from annual or biannual crops do not need to occupy a field for that length of time, in principle, and can be grown in rotation with other crops depending on soil conditions. Companies may still prefer longer occupancy when having to invest in soil preparation and infrastructure (see Table 5.1).

5.4 Social impacts

Social impacts of biofuel cropping systems depend on what benefits are generated and the extent to which risks are mitigated. Benefits include the following: more opportunities for local enterprises (mostly farming smallholders) – such as improved access to inputs, technologies, and markets; creation of employment that is accessible to the local population who have no or limited formal education; and other multiplier effects on the economy. Risks include the following: loss of access to resources; displacement without compensation, resulting in loss of livelihood; arrival of large numbers of migrants; upheaval of existing governance and community structures; and environmental destruction. Gender analysis is also important to reduce the risk of most benefits going to men while women have to shoulder most of the negative impacts. The results will be influenced by the engagement in negotiations and bargaining powers of communities and organisations of

Table 5.1 Land-use requirement per biofuel cropping system

	Sugarcane	Oil Palm	Soybean	Cassava	Jatropha	Tree Crops
Tenure system characteristics						
Duration land occupation	Permanent: 5–15 years	Permanent: 25 years	Annual crop	18 months	Permanent: 40 years	Permanent: >20 years
Water requirements	High: Often grown in former wetlands	High: Limited irrigation	High	High: Drought-resistant, but affect productivity	Medium–High: Drought-resistant, but affect productivity	Variable
Reported land-use conversion	Wetlands, forests	Forests	Forests, grazing lands	Grazing lands	Grazing lands, forests	Natural forests
Reported tenure conflicts with smallholders	Yes	Yes	Yes	No	Yes	Yes

smallholders and workers, along with the government's role in protecting the rights of smallholders and workers. The ability to negotiate requires information and capacity, and communities are likely to require external legal support from organisations with more experience in these types of negotiations. Equally, when communities receive some form of compensation, transparent systems of accounting for these resources and solid planning for investment are required to ensure that all community members benefit and will be compensated for their losses.

Important factors for biomass-producing companies to improve economic and social outcomes and reduce risks are: (1) the business model chosen for producing feedstock and whether this involves smallholders and other local enterprises; (2) the type of contract relations between producers and processors; (3) the type of employment generated, the training offered, and the recruitment process; (4) benefit-sharing systems (shareholding and other revenue sharing); (5) tax payments to local governments and support to mechanisms to ensure accountability; and (6) wider spin-off and societal effects of investments in infrastructure and social services in the enterprise, such as health, education, housing, or roads. Other indirect impacts depend on the multiplier effects of the enterprise around the supply chain and in the area. Crops that can be used both for food and for biofuel may offer more opportunities.

The models that engage smallholder farmers will have the highest socio-economic effects and are likely to improve community relations and create less reputational risk for companies around land acquisition. These models will be more demanding in organisation and management. Collaboration with experienced local organisations may reduce these challenges. Smallholders may benefit if they can supply their produce with terms that are beneficial to them. Outgrower schemes and contract farming are not necessarily positive or disadvantageous to smallholders. The impact depends on the division of costs, benefits, and risks. This in turn depends on whether the interests of smallholders are effectively represented in the contract negotiations (see Table 5.2) (Baumann, 2010; Vermeulen and Cotula, 2010).

5.5 Discussion and conclusion

This final section discusses some entry points. An understanding of which investments will have multiplier effects, what the role of government should be, appropriate planning, and clear guidance to companies will generate mutually beneficial outcomes.

One reason for governments to welcome investors is that governments cannot fund investment in infrastructure themselves. Public benefits will be limited, though, if companies' *investments* in infrastructure tend to be concentrated first on their own farms. They may even expect governments to supply roads, electricity, and telecommunication services. In irrigated areas, investors tend to avoid major investments by acquiring fields next to existing

Table 5.2 Social impact of various biomass crops

	Sugarcane	Oil Palm	Soybean	Cassava	Jatropha	Tree Crops
Cropping system characteristics						
Multiple uses including food crops	Yes	Yes	Yes	Yes	No	No
Local processing industry	Food, juice, alcohol	Food (West Africa)	Food (NGO driven)	Food	Soap	Timber, charcoal
Possibility to cultivate on small plots (cost of production/price)	Yes, requires link with factory	Yes, requires link with factory	Yes	Yes	Yes	No data
Already cultivated by smallholders	Yes	Yes	Yes	Yes	Yes (hedges)	No data
Smallholders (outgrowers) experience for biofuel	Yes	Yes	Yes (Brazil)	No data	Yes	No data
Employment opportunities on plantations	High	Medium	Low	No data	Low to medium	Low
Potential local economic development	Depends on wages, land rent	Depends on wages, land rent, local processing industry	Limited	No data	Depends on wages, land rent, nut prices	Limited

infrastructures, preferably more upstream, which will enhance competition with existing farmers (see earlier in this chapter). Costs and benefits of infrastructure investments have to be clarified in the contract.

Tax and other revenues for government

A source of income for local or national governments, they can be invested in economic development or basic service provisions. Whether biomass production systems generate revenues and taxes depends on the conditions in the contracts and their actual enforcement. Low valuation of the land, generous terms of the lease, and tax holidays will not generate much benefit for the government. Elements of importance are the land rent charged, whether it is actually collected, whether tax exemptions are granted, and also whether the government has to invest in roads, electricity, and so on.

Protect ecosystems services

The expansion of biomass production may cause land-use changes such as clearing of primary forest area, planting of wetlands, or ploughing of grazing areas. The effect on wider ecosystem services, such as watersheds or soils, may affect even the environmental sustainability of biomass farms. The impact depends on the fragility of the ecosystems (e.g., biodiversity, wildlife routes, soil type, slope) and the function of the area in hydrological systems. Environmental impact assessments should, in theory, create safeguards. In practice, however, there is a tendency by governments not to comply, to relax, or even to lift restrictions to enable investments. In addition, the lifting of the status of "protected area" to enable expansion of biomass production will not serve ecosystems. In Indonesia, for example, oil palm first led to deforestation and is now causing the loss of peat lands, which are both less productive and ecologically more vulnerable (Susanti and Burgers, 2011). This race to the bottom with respect to environmental regulations is not in the interest of sustainable production systems, and companies that have a long-term outlook could indicate to governments that such policies may even discourage investments. There are initiatives to orient the expansion of oil palm towards "degraded" farmland rather than forests (Austin et al., 2012).

Governance

The way that deals have been concluded with respect to making land available for biomass production is important for governance in general, and also for the business climate of companies that aim for long-term engagement. One challenge is that local rights are not sufficiently protected (Cotula, 2011; HLPE, 2011; Oxfam International, 2011). In addition, the contracting process may involve tricky governance challenges for those companies involved when there is limited transparency in the land allocation process or over the contract conditions; all arrangements remain confidential, and the national parliament may not be in a position to scrutinise the contracts. Clearly, transactions shrouded in secrecy with limited transparency and accountability with respect to contract negotiations with relevant government agencies, local government, parliament, or communities will not contribute to improving governance. Companies often have to deal with muddy waters and are faced with a range of actors and interests – governments, competitors, finance agencies, civil society organisations, drivers positions in communities, and so on – and have to act at national and local levels. Therefore, the actions of companies and their choice of business model – protection of the environment, contract transactions, and due diligence – matter for the outcome of their socioeconomic activities.

Voluntary principles and standards for the EU-based private sector

A number of these standards concern land-based investments. These standards are generally voluntary but can become mandatory when the company is from an OECD country or is requesting funding from the IFC or from development funds that follow the standards set by the IFC, or if the company intends to participate in roundtables for sustainable production (oil palm, soy, or biofuels). The standards set influence possibilities for import into Europe and the USA, with supply chain management gaining in importance,[1] as well as avoiding reputational risk. Increasingly, companies have to demonstrate that production is done in a 'responsible' way, including respect for land rights.

The IFC guidelines are being reviewed to take better account of issues related to food security and respect for land rights. Pension funds have recently launched guidelines on farmland principles.

Biomass production can be undertaken in such a way that positive socioeconomic effects are maximised, environmental impacts are reduced, and risks for local communities are mitigated. This requires explicit strategic choices by the companies, building on experience available in the sector or in other sectors with land-based investments (e.g., mining); with similar types of enterprises and a multistakeholder approach, such as working with organisations that know the area; with local governments; and so on. The costs involved are offset by access to more produce (if working with smallholders), improved community relations, and reduced risks such as with respect to conflict. Private-sector actors play an important role in the outcomes, and companies can lead by example and show that it is possible to develop a profitable and sustainable enterprise while respecting local rights, contributing to economic spinoff, and minimising environmental degradation.

Note

1. See the Wilmar palm oil company case in Indonesia. Following a complaint by NGOs in 2009 that agreed procedures for sustainable palm oil were not adhered to, the IFC suspended its loan to Wilmar; international processors no longer bought the palm oil.

References

Alden Wily, L. (2010). *Who owns Africa? Looking through the lens of community based tenure.* Washington, DC: Rights and Resources.

Austin, K., Sheppard, S., and Stolle, F. (2012). *Indonesia's Moratorium on New Forest Concessions: Key Findings and Next Steps.* Washington DC, USA: World Resources Institute. Retrieved on 10 November 2013 from: http://www.wri.org/publication/indonesia-moratorium-on-new-forest-concessions

Baumann, P. (2000). *Equity and efficiency in contract farming schemes: The experience of agricultural tree crops.* London: ODI.

Baumgart, J. (2011). *Assessing the contractual arrangements for large scale land acquisitions in Mali with special regard to water rights.* Presented at the Annual World Bank Conference on Land and Poverty, April, Washington, DC.

Cotula, L. (2011). *Land deals in Africa: What is in the contracts?* London: IIED.

de Schutter, O. (2011). The green rush: The global race for farmland and the rights of land users. *Harvard International Law Journal*, Vol. 52, pp504–559.

HLPE. (2011). *Land tenure and international investments in agriculture.* A report by the High Level Panel of Experts on Food Security and Nutrition of the Committee on World Food Security, Rome.

Morris, M., Binswanger-Mkhize, H.P., and Byerlee, D. (2009). *Awakening Africa's sleeping giant: Prospects for commercial agriculture in the Guinea savannah zone and beyond.* Washington, DD: World Bank Publications.

Palmer, D., Fricska, S., and Wehrmann, B. (2009). Towards improved land governance. Land tenure working paper no.11. Rome, Italy: FAO.

Susanti, A., and Burgers, P. (2011). *Oil palm expansion: Competing claim of lands for food, biofuels and conservation.* Paper presented at ICCAFFE Conference 19–21 May, University of Agadir, Morocco.

Vermeulen, S. and Cotula, L. (2010). *Making the most of agricultural investment: a survey of business models that provide opportunities for smallholders.* IIED/FAO/IFAD/SDC, London/Rome/Bern.

Oxfam International. (2011). *Land and power: The growing scandal surrounding the new wave of investments in land.* Oxfam Briefing Paper 151. Oxford: Oxfam International. Retrieved from www.oxfam.org/grow

6 Biofuel production in Brazil

J.W.A. Langeveld and P.M.F. Quist-Wessel

6.1 Introduction

Brazil, the largest country in Latin America and the world's seventh-wealthiest nation (in 2011), has a GDP of more than US$2 trillion and a population of 190 million. This former developing country reached middle-income status and now is considered a dynamic emerging economy similar to Russia, India, China and South Africa (the so-called BRICS countries). Since 2001 sustained economic growth, low inflation, social programs, and increased minimum wage have helped to halve poverty. Between 2004 and 2009 the share of the population earning less than US$1.25 per day (the extremely poor) dropped from 10% to 2%, which is an amazing achievement. During that period, the poor enjoyed income growth rates four times higher than those of the rich. Despite this, inequality remains high, and there is still a large gap between the poor and rich; for example, with respect to education (World Bank, 2013).

Brazil is the world's second-largest producer of ethanol (after the USA) and has been the pioneer in developing integrated biofuel policies and infrastructure. Biofuel production and consumption are integrated in the economy to a level not realised elsewhere, making Brazil a model both for developed and developing countries. Ethanol fuel is available throughout the country, both as a blend with gasoline and as a separate fuel (E85). This success is based on a combination of factors, including several decades of a support program, extensive land availability, effective agricultural knowledge, and strong industrial development. Dedicated research has helped Brazil to develop the most successful alternative fuel to date. Its energy efficiency and reduction of greenhouse gases (GHG) has been rewarded the status of 'advanced biofuel' by the US Environmental Protection Agency (EPA).

The history of ethanol in Brazil is unique and serves as a successful example of biofuel development. Its history and background will be discussed in this chapter, which is organized as follows. First, background data are presented on land availability and land use in Brazil (Section 6.2). Next, market development and economic conditions for biofuel production are discussed (Section 6.3), after which practices of crop cultivation and biomass production are explained (Section 6.5). This is followed by an evaluation of the impacts of biofuel production (Section 6.7) and some conclusions (Section 6.8).

6.2 Land resources

Brazil, covering 850 million ha, is nearly the size of the USA. It consists mostly (62%) of forest and agricultural land (31%). The share of 'other' land, 60 million ha or 7%, is relatively low. Figure 6.1 depicts a schematic overview of different types of land use in Brazil.

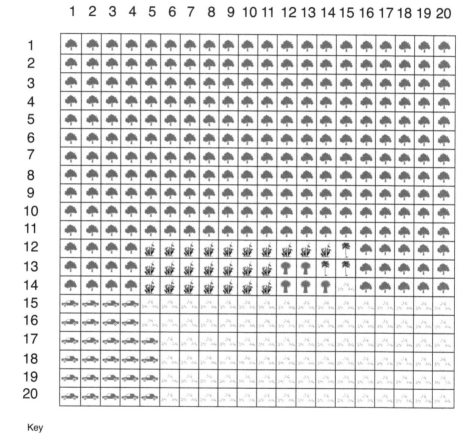

Key

🌲 forest   permanent grassland   arable crops

fodder   agricultural tree crops   other

Figure 6.1 Schematic overview of land use in Brazil

No geographical representation. Each cell represents 0.25% of land area. Position of the categories was chosen randomly.

Source: Calculated from FAOSTAT (2010–2013), http://faostat.fao.org

Table 6.1 Land area and land use in Brazil

	Unit	1980	1990	2000	2010
Agricultural area	million ha	224	242	261	273
Permanent grassland	million ha	171	184	196	196
Agricultural tree crops	million ha	8	7	8	7
Arable area	million ha	45	51	58	70
Of which					
arable crop area	million ha	46	42	40	51
fodder crop area	million ha	No data	9	18	12
Fallow	million ha	No data	No data	No data	8
Forest area	million ha	No data	575	546	520
Other land	million ha	No data	30	39	53
Multiple Cropping Index (MCI)	–	1.06	0.88	0.74	0.85

Sources: Calculated from FAOSTAT (2010–2013), http://faostat.fao.org; Siebert et al. (2010).

Brazil has 273 million ha of agricultural land (Table 6.1). Three-quarters of this is permanent grassland. Agricultural tree crops (fruits, coffee, oil palm) cover 3 7 million ha. The remainder (70 million ha) is mostly used for arable crops including cereals, pulses, oil crops, and sugarcane. Fodder crop area is estimated at 12 million ha (17% of arable area). Estimates of fallow are extremely rare. Following Siebert et al. (2010), we assumed 10% of arable land under fallow. The share of forest land, although declining, still is very high.

Brazil has tropical conditions in the north and a temperate climate in the south. Temperatures around the equator are high, averaging above 25°C. Winters (June–September) may bring frost and snow in states south of the Tropic of Capricorn. Most of Brazil has moderate rainfall ranging between 1,000 and 1,500 mm a year, peaking in the summer months between December and April. The Amazon region is very humid with rainfall in excess of 2,000 mm, but it has a distinctive dry season of from three to five months. The northeast has erratic rainfall and severe droughts, and constitutes the highest temperatures with a seven-month dry season (May–November). Rainfall in the centre west is 1,500–2,000 mm, with a pronounced dry season in the middle of the year ('Climate of Brazil', 2013).

Brazil is dominated by low activity clays with small inclusions of other soil types. Soils of the Cerrado in the south-central region are highly acidic, saturated with aluminium and deficient in phosphorous, and they have low water-holding capacity. It was long felt that these soils could not be cultivated, but they proved to have excellent physical characteristics suitable for mechanised crop production (e.g., deep and well drained), while improved fertilisation regimes allowed successful cultivation. Half of the Cerrado area, or 95 million ha, appeared to be suitable for large-scale crop production.

Soils in the more northern production regions tend to be low in clay but do not suffer from chemical restrictions such as aluminium toxicity.

Agriculture is characterised by large estates alongside a large number of small farms. In 1985 some 90% of the 6 million agricultural holdings held less than 100 ha. They generated a considerable share of crop production, varying from 14% of the sugarcane to 85% of cassava. Small farm production in soybean is 37%. The contribution of small farmers in some regions is below the national average. This is the case for sugarcane production in the southeast, northeast, and south-central regions and for soybeans in the centre west region. Large farms are concentrated mainly in the centre west, where soybeans predominate as the main crop (FAO, 2004).

There are large differences with respect to cultivation practices. Limited adoption of productive technologies explains low crop productivity in the north and northeast regions. Advisers from the extension service, for example, visited only 4–7% of farms in the north and northeast. In those regions, only 10–18% of the farmers used lime and fertilisers while less than 1% practiced some kind of soil conservation. Irrigation was not common; only 5% of the farms had been applying irrigation in the dry northeast (FAO, 2004).

6.3 Markets and policies

As one of the leading nations on climate negotiations, Brazil has committed itself to reducing its GHG emissions by 36% until 2020 (World Bank, 2013). This is in line with a strong and successful implementation of renewable energy and transport fuels, but biofuel policies were originally developed to save on expensive fossil fuel imports. Proálcool, a stimulation programme for ethanol production, was implemented immediately after the first oil crisis in 1974, which pushed oil prices to unprecedented levels (Table 6.2). The programme was based on three drivers: (1) guaranteed purchases by the (state-owned) oil company, (2) low-interest loans for ethanol firms, and (3) fixed gasoline and ethanol prices. Ethanol was sold at the pump for a fixed price level set at two-thirds of the government-set gasoline price ('Ethanol Fuel in Brazil', 2012), which is in line with differences in energy contents.

Proálcool had to address many technical issues while occasionally demand for ethanol plunged when oil prices were low. Presently, light vehicles in Brazil are no longer running on pure gasoline. Since blending became mandatory, blending shares fluctuated between 10% and 25%. In 1993 the mandatory blend was fixed at 22% (E22) by volume. Since 2007 the mandatory blend has been 25% of anhydrous ethanol and 75% gasoline or E25 blend. The lower limit was reduced to 18% in April 2011 due to recurring ethanol supply shortages and high prices that take place between harvest seasons ('Ethanol Fuel in Brazil', 2012).

In 2010 exceptionally high sugar prices made it more profitable for sugar mills to produce sugar, and sales of hydrous ethanol fell due to relatively

Table 6.2 Thirty years of biofuel policy and production in Brazil

Year	Event	Biofuels
1973	First petroleum crisis	
1974		Launch of the Proálcool ethanol program
1977		4.5% ethanol added to gasoline
1979		15% ethanol added to gasoline
1980	Second petroleum crisis	
1985		Mandatory blend of 22% ethanol
1989	Drop of petroleum prices	Gasoline price equals ethanol price
1993		Mandatory blend of 20–25% ethanol
2003	Introduction of flex-fuel technology	
2005	1.4 million flex-fuel cars sold	Launch of PNPB biodiesel program including mandates and tax exemptions
2007	Third petroleum crisis	
2008		Ethanol equals gasoline consumption; mandatory blend of 2% biodiesel
2010		Mandatory blend of 5% biodiesel
2011		Ethanol crisis in Brazil; ANP and Ethanol (18–25%)
2012		Launch of PNPB phase 2

Sources: Souza (2012); Valdes (2011); 'Brazil' (2012).

high prices at the pump. Despite the reduction in ethanol consumption, total sales reached 22 billion litres while pure gasoline consumption was 23 billion litres, keeping the market share close to 50%. Presently, anhydrous ('dry') ethanol is added to gasoline at a rate of 25%. Hydrous ethanol is also sold for use in flex-fuel cars that handle any blend of ethanol and gasoline.

The first biodiesel programme, starting with vegetable oil–based fuels in the 1940s, was abandoned in 1984. It was reinstalled in 2005 with the launch of a National Program for Production and Use of Biodiesel (PNPB), which aimed to stimulate fuel diversification, social inclusion, and regional development. These objectives were to be realised by adding value to the soy production chain, which was suffering from a surplus of soy oil, originally a by-product of protein production.

The biodiesel program mandated a 2% blend of biodiesel (B2) in 2008, requiring 1 billion litres of biodiesel. The target was raised to 5% (B5) in 2010, requiring 2.4 billion litres of biodiesel. Federal tax exemptions were imposed in 2005 for fuel producers who were using feedstocks provided by small farmers in given regions while providing access to cheap credit. Soybean biodiesel was excluded from tax exemption (De Almeida et al., 2007).

Biodiesel production capacity has been installed at a high rate. By the end of 2007, installed capacity amounted to 1.5 billion litres, with another 2 billion litres under construction, and a further 2 billion litres projected (Gazzoni, 2007). In 2009 the capacity was estimated at 4 billion litres while nearly 6 billion litres was planned for the near future (De Almeida et al., 2007). One year later, a total of 68 refineries had a cumulative production capacity exceeding 6 billion litres (Talamini and Dewes, 2013).

An overview of biofuel production is given in Table 6.3. Since 2000 cane ethanol production has reached a level of 28 billion litres. Domestic consumption doubled to 22 billion litres in 2010. Initially, the ethanol sector received important financial support. According to Steenblik (2007), huge loans that were given were often not repaid. Over a period of 30 years, Proálcool investments totalled some US$11 billion, which is very modest. Presently, however, ethanol production has become mature and economic.

Ethanol production costs are around US$0.22 per litre, following a steady decline due to technological development and up-scaling, and production currently is competitive without subsidies (Sanchez and Cardona, 2008). In general, cane ethanol is competitive with gasoline at fossil oil prices of US$40–50 per barrel or higher (C2ES, 2012), and import tariffs have in practice been cut to zero (Valdes, 2011). As Brazilian gasoline taxes are high (54%), while ethanol taxes are between 12% and 30%, production of ethanol could also be competitive at lower oil prices ('Ethanol Fuel in Brazil', 2012).

In 2008 most biodiesel plants intended to use a range of feedstocks, though the most important by far is soy oil (De Almeida et al., 2007). Some plants were not operational due to high vegetable oil prices in 2007 and 2008 or to switching to cheap sources such as tallow. Overcapacity and high feedstock prices appeared to cause a shake-out, leading to scale enlargement.

Table 6.3 Market conditions and biofuel production in Brazil

		Bioethanol		Biodiesel	
	Unit[1]	2000	2010	2000	2010
Production	billion litre/year	9.7	27.6	Negligible	2.1
Consumption	billion litre/year	13.7	22.2	Negligible	2.7
Support investments	billion US$	0.4[2]			
Import levies	US$/litre		0		14%
Production cost	US$/litre	0.20	0.22		

[1]Billion = thousand million
[2]Averaged figure for 1970–2000

Sources: Calculated from OECD (2008); OECD-FAO (2012); FAPRI-ISU (2011); Sanchez and Cardona (2008); Valdes (2011); 'Ethanol Fuel in Brazil' (2012)

With some 80% of the biodiesel cost being related to feedstocks, it makes biodiesel very dependent on commodity market prices.

Further increases of biofuel production have been projected for 2020 – ethanol production possibly expanding to 48–55 billion litres while more than 3 billion litres of biodiesel could be generated (FAPRI-ISU, 2011; OECD-FAO, 2012). Meanwhile, the government aims to raise the levels of the addition of biodiesel to diesel in the near future. With new biodiesel plants under construction, production capacity could be doubled (Talamini and Dewes, 2013).

6.4 Feedstock requirements

Biofuel production is dominated by sugarcane (ethanol) and soybean (biodiesel) (FAPRI-ISU, 2011). Soybean biodiesel remains the main feedstock, as dedicated feedstock sources such as oil palm and Jatropha (and maybe castor) need time to develop (De Almeida et al., 2007).

Sugarcane for ethanol

Sugarcane is a tropical grass with the effective C4 photosynthetic pathway with a great production potential.[1] Although it can produce flowers and seeds, it is harvested before this occurs in order to ensure a good crop. Brazil has two major regions for sugarcane production: the dry northeast and the more humid south-central region (Figure 6.2).

Nearly 90% of the crop is found in the south-central region, which hosts the Cerrado, an important cane area. Agricultural development in the Cerrado started in the 1970s when major limiting factors to crop production were addressed. A research programme was initiated to correct phosphate and soil acidity problems, and good phosphate recoveries now can be realised provided that soil pH is increased to 6.0–6.5 and exchangeable aluminium is reduced. Consequently, this previously unproductive region currently is providing half of the country's maize, soybean, and beef (FAO, 2008a).

Sugarcane area, production, and yield are presented in Table 6.4. Cultivation quadrupled since 1980 to a net 9 million ha. Area expansion and production have accelerated since 2000. Yield improved from 56 tonne/ha in 1980 to almost 80 tonne/ha in 2010. Average annual yield improvement in the 2000s (based on three-year average yield figures) amounted to more than 1,200 kg per ha per year.

Cane use for biofuels doubled between 2000 and 2010, when nearly 350 million tonnes was converted (Table 6.5). It is projected to double again over the next decennium. Half of the harvest is devoted to ethanol, although in practice, world market prices for sugar and ethanol steer the actual division between alcohol and sugar.

Table 6.4 Sugarcane production in Brazil

	Unit	1980	1990	2000	2010
Harvested area	million ha	2.6	4.3	4.8	9.1
Production	million tonne	149	263	328	717
Yield (three-year average)	tonne/ha	55.6	61.8	68.5	79.5
Average annual increase[1]	kg/ha/year		617	670	1,233

[1]Figure for 1980–1990 presented in the 1990 column; figure for 1990–2000 in the 2000 column, and so on.

Source: Calculated from FAOSTAT (2010–2013), http://faostat.fao.org.

Table 6.5 Sugarcane use for ethanol production

	Unit	2000	2010	2020
Use in biofuels	million tonne	170	348	647
Share of all cane	%	52%	48%	63%
Biofuel area	million ha	2.5	4.3	7.4

Sources: Calculated from FAPRI-ISU (2011); FAOSTAT (2010–2013), http://faostat.fao.org; Goldemberg and Moreira (1999).

Soybean for biodiesel

Soybean (USA), or soya bean (UK) (Glycine max), is a leguminous crop originating from China. It is an annual C3 plant that produces beans and oil seeds. The leaves fall before the seeds mature. Seeds are rich in proteins (38–43%), oil (18–25%), and carbohydrates (24%) (El Bassam, 2010).

Agronomic research in the 1970s developed soybean varieties fit for cultivation in low-latitude regions, a feat that is considered one of the most significant technological innovations in the Green Revolution. New fast-growing ('short-cycle') varieties facilitated the application of double cropping (two crops cultivated sequentially on the same plot of land), while no-till planting (not requiring ploughing during land preparation) reduced the risk of soil degradation (USITC, 2012).

Research resulted in the development of high-yielding soybean varieties tolerant to high aluminium soils found in the Cerrado and dry soil conditions. As a result, present soybean varieties grown in Mato Grosso yield slightly higher than their counterparts grown in the USA (McVey et al., 2000). Five states are responsible for 80% of Brazil's soybean production: Rio Grande do Sul and Paraná in the south (older cultivation areas) and Mato Grosso, Goîas, and Mato Grosso do Sul in the centre west (Cerrado region).

An overview of soybean production is given in Table 6.6. The soybean area increased strongly from 1980 to 2000, and it has doubled since 2000. This increase exceeds that of sugarcane and also exceeds the expansion of

Table 6.6 Soybean production in Brazil

	Unit	1980	1990	2000	2010
Harvested area	million ha	8.8	11.5	13.6	23.3
Production	million tonne	15	20	33	69
Yield (three-year average)	tonne/ha	1.6	1.8	2.5	2.8
Average annual increase[1]	kg/ha/year		17	77	31

[1]Figure for 1980–1990 presented in the 1990 column; figure for 1990–2000 in the 2000 column, and so on.

Source: Calculated from FAOSTAT (2010–2013), http://faostat.fao.org.

Table 6.7 Soybean use in biodiesel production in Brazil 2000–2010

	Unit	2000	2007	2010	2020
Use in biofuels	million tonne	Negligible	2	9	11
Share of all soybean	%	Negligible	3%	14%	13%
Biofuel feedstock area	million ha	Negligible	0.1	0.4	0.4

Source: Calculated from FAOSTAT (2010–2013), http://faostat.fao.org; FAPRI-ISU (2011).

agricultural or arable area, suggesting that soybeans increasingly are replacing other crops. Yield improvement has been variable; it was very high in the 1990s (average annual increase amounting to nearly 80 kg/ha/year) but declined after 2000.

The use of soybeans in the biodiesel industry quadrupled within a few years, starting at 1.6 million tonnes in 2007 to reach 11 million tonnes in 2010 (Table 6.7). It is projected to increase only slightly towards 2020. Notwithstanding recent increases, a modest part of soybean production is allocated to biofuels. The share increased from 3% in 2007 to 14% in 2010 and is projected to remain at this level.

6.5 Crop cultivation

Sugarcane

The optimum temperature range for sugarcane cultivation is 25–30°C. High temperatures, high humidity, and moist soil provide favourable conditions for crop growth, but a period with cool dry weather is needed to promote ripening (IFA, 2012). The crop is harvested after a minimum of 10 months and a maximum of 24 months, mostly during the cooler and relatively dry part of the year (Figure 6.2).

Harvesting and planting operations tend to overlap in order to avoid the need to store planting material which has to be used fresh (IFA, 2012).

Activity	Jan	Feb	Mar	Apr	May	Jun	Jul	Aug	Sep	Oct	Nov	Dec
Planting	X	X	X	X					X	X	X	X
Growing season	X	X	X	X	X	X	X	X	X	X	X	X
Fertilising				X	X	X	X	X	X	X	X	
Harvesting				X	X	X	X	X	X	X	X	
Post-harvest	X	X	X	X	X	X	X	X				X

Figure 6.2 Sugar cane cropping calendar in Brazil

Sugarcane is harvested annually for five or six consecutive years. During the harvest season, running from April to October in the Cerrado region, generally 20% of the sugarcane is removed and replaced with other crops such as beans, peanuts, or corn for soil improvement during one season.

Sugarcane requires medium to heavy to soils with a pH ranging from 5.0 to 8.5. Liming is required if the pH is lower than 5 (e.g., in the Cerrado region), or gypsum if the pH is higher than 9.5. Plants are reasonably tolerant to waterlogging for up to 7–10 days, except during tillering and sprouting (IFA, 2012).

Sugarcane is not a very demanding crop (Table 6.8). Nutrient applications reported by FAO (2004) are modest at 68 kg of nitrogen, and 45 kg of phosphate and 130 kg of K_2O. Application levels reported by BioGrace (2012)

Table 6.8 Sugarcane cropping in Brazil: input requirements

Parameter (unit)	Per ha	Source
Nutrient application (kg)		
N	68	FAO (2004)
	63	BioGrace (2012)
P_2O_5	45	FAO (2004)
	28	BioGrace (2012)
K_2O	130	FAO (2004)
	74	BioGrace (2012)
Lime	600	Do Amaral et al. (2008)
Application of agro-chemicals (kg active ingredient)		
Herbicides	0.22	Goldemberg et al. (2008)
Insecticides	0.04	Goldemberg et al. (2008)
Other pesticides	0.11	Goldemberg et al. (2008)
Diesel use (litre/ha)	55	BioGrace (2012)
Required rainfall (mm)	1,500–2,500	Goldemberg et al. (2008)

are similar for nitrogen but show important deviations for phosphorus and potassium (BioGrace reporting 50% lower applications). Lime is needed to combat soil acidity. Considerable amounts of phosphate and potassium can be recovered by recycling vinasse and filter cake from sugar and ethanol production facilities.

Application of agro-chemicals (mostly herbicides) is low compared to that of other arable crops (maize, soybean) but shows large variations over cropping systems. Diesel use is very modest. Water requirements are rather high but can be supported by precipitation in most production regions. Irrigation is not common.

Sugarcane used to be burned before (manual) harvesting in order to get rid of leaves which form a risk for labourers. According to Sawyer (2008), 80% of the fields were burned in 2008. This system is gradually being replaced by mechanical green harvesting where burning is no longer needed.

Soybean

Soybean is cultivated in regions where temperatures during the planting are between 20°C and 30°C (Gazzoni et al., 2002). Flowering is induced by low (below 13°C) temperatures (Embrapa, 2010). A precipitation level of 450–800 mm is required during the growing season, but senescence and harvest must coincide with low rainfall periods to ensure grain quality (Gazzoni et al., 2002). Drought can strongly reduce yield, especially in the period before flower formation or grain forming (El Bassam, 2010).

Seeds in the main soy area are planted in September or October. Soybean harvest starts in December in the south and January or February further north. The crop is kept free from weeds during the first 45–50 days after emergence through mechanical (hoe or field cultivator), chemical, and cultural practices. Mechanical harvesting requires a plant height of about 60 cm (Embrapa, 2010). A typical crop calendar is presented in Figure 6.3.

Soybean is grown as a mono-crop or in rotation with maize, wheat, or other crops. Rotations are more common in southern states as compared to central and northern states. Zero-tillage systems have been rapidly adopted

Activity	Jan	Feb	Mar	Apr	May	Jun	Jul	Aug	Sep	Oct	Nov	Dec
Preparation									X	X		
Sowing									X	X	X	
Growing season	X	X	X							X	X	X
Harvesting	X	X	X									
Post-harvest				X	X							

Figure 6.3 Cropping calendar for soybean cultivation in Brazil

from the 1980s onwards in response to severe soil erosion problems (Goedert, 1983), allowing farmers to reduce the demand for machinery, diesel, and labour. In some cases, soybean is followed by a relay crop (e.g., corn).

Soybean requires well-drained soils with adequate water retention capacity and fertility. Ideally, soils should be of medium texture (clay content from 30% to 35%) or should be heavy but well drained. Soils with low (<15%) clay content are to be avoided, and the crop is not to be sown on slopes exceeding 8% when using conventional cropping systems (or over 15% when using the no-till systems). Soil depth should at least be one metre, and major physical problems, such as the presence of stones or salty soils, should be avoided (Gazzoni et al., 2002).

Soybean is a leguminous crop that fixes nitrogen from the air through a symbiotic process with rhizobium bacteria. Brazil has a program for selection of effective rhizobia strains. This program has been highly successful in reducing the need for nitrogen fertilisation. For the production of 1,000 kg seeds, soybean requires 20 kg P_2O_5/ha and 20 kg of K_2O (Embrapa, 2010). According to IFA (2012), there is no need for nitrogenous fertilisers, but in practice, small applications occur.

An overview of input used is presented in Table 6.9. Application rates of phosphate and potassium fertilisers depend on the nutrient status of the soil. In practice, some 30 kg of P_2O_5 and K_2O is given, which indicates that soil fertility generally is good. Application of agro-chemicals is very high; especially the need for pesticides and herbicides. Total consumption is over 11 kg of active ingredient per ha. Consumption of fossil fuels during cultivation (59 litre per ha) is modest.

Table 6.9 Soybean cultivation in Brazil: input requirements

Parameter (unit)	Per ha	Source
Nutrient application (kg)		
N	8	FAO (2004)
	66	FAO (2004)
P_2O_5	30	Common practice
	62	FAO (2004)
K_2O	30	Common practice
Application of agro-chemicals (kg active ingredient)		
Pesticide	6.6	Meyer and Cederberg (2010)
Herbicides	4.2	Meyer and Cederberg (2010)
Fungicide	0.6	Meyer and Cederberg (2010)
Insecticides	1.0	Meyer and Cederberg (2010)
Diesel use (litre/ha)	59	BioGrace (2012)
Required rainfall (mm)	450–800	Embrapa (2010)

Optimum plant population and row spacing depend on soil texture and fertility, climatic conditions, and prevalence of pests. The sowing depth is 3–5 cm, and planting density varies from 200,000 to 400,000 plants per ha. In general, cultivars that are tall and/or have a long crop cycle require lower densities (Embrapa, 2010).

Most soybean varieties in Brazil are genetically modified (GM), with the share of GM soybean increasing from 2% to over 60% between 1997 and 2007. The so-called Roundup Ready gene, which brings tolerance to the effective but nonselective herbicide type of glyphosate, was incorporated in a wide range of varieties adapted to local conditions, thus allowing GM cultivation in a range of conditions.

About two-thirds of soybean operations in centre-west Brazil cover more than 2,500 ha, and many are more than 20,000 ha. These mega farms were formed when huge undeveloped tracts of land became available in the Cerrado just as improved technology allowed improved yields and mechanised harvests on flat parcels in tropical areas. Southern Brazilian soybean farms are much smaller – in Paraná most farms are less than 60 ha (USITC, 2012).

6.6 Conversion to biofuels

Biofuel production in Brazil is dominated by sugarcane and soybean.

Sugarcane to ethanol

An overview of the cane-to-ethanol conversion process is presented in Figure 6.4. Sugars make up one-third of the crop, with the remainder being fibres generally referred to as bagasse. Cane stalks are crushed to yield juice and bagasse. The latter is burnt to cover the energy requirements for ethanol production. Energy surpluses are used to generate electricity (co-generation). The cane juice is filtered, purified, and subjected to a fermentation process after which ethanol is won through distillation. The favoured microorganism in cane ethanol production, *Saccharomyces cerevisiae,* has a good capability to hydrolyse sucrose into glucose and fructose (Sanchez and Cardona, 2008).

Cane conversion is done in batches or in continuous processes. Fed-batch culture is the most common fermentation technology in Brazil. Control of medium flow rate helps to prevent inhibitory effects of high substrate or product concentrations in the fermentation broth. This can lead to increases of ethanol productivity of 10–14%. Continuous processes are run in relatively cheap bioreactors. These allow better process control, require less maintenance, and are cheaper to operate. One-third of the production facilities employ continuous fermentation (Sanchez and Cardona, 2008).

Brazil hosted 423 plants in 2009, half of which produced sugar alongside ethanol, while 159 produced just ethanol. All mills produce their own

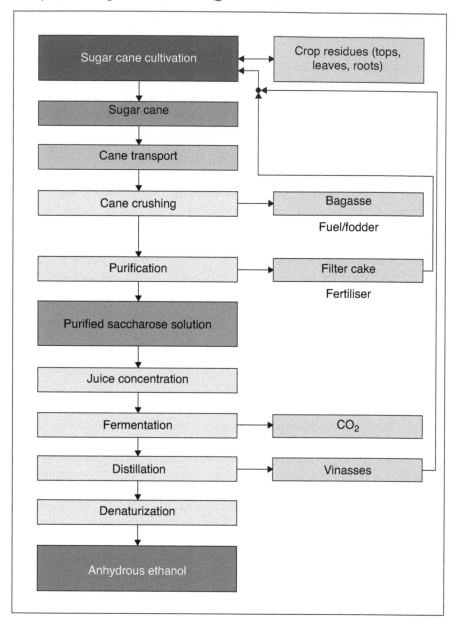

Figure 6.4 Conversion of sugarcane to ethanol. Figure design by H. Croezen.

electricity (Neves et al., 2010). Over the past three decades, cane-to-ethanol conversion improved from 70 to 100 litres per tonne (Table 6.10). Some 36% of cane energy is recovered in ethanol (BioGrace, 2012). Following BioGrace (2012), 1 ha of cane yields 5,000 litres of ethanol. Other estimates place yields as high as 7,900 litres.

Table 6.10 Efficiency of sugarcane-to-ethanol conversion in Brazil

	Unit	BioGrace	Other Sources
Conversion efficiency	litre/tonne	100	70–100
Conversion efficiency	GJ/GJ	0.36	
Biofuel yield	litre/ha	5,000	7,900
Biofuel yield	GJ/ha	134	

Sources: Calculated from BioGrace (2012); FAO (2008b); Burrell (2010); Sanchez and Cardona (2008).

Because there are no relevant co-products that are used outside the cane ethanol production chain, all inputs used during cultivation, transport, conversion, and distribution are allocated to the ethanol (BioGrace, 2012).

Soybean to biodiesel

Conversion for biodiesel production from soybean is depicted in Figure 6.5. The process consists of three steps. First, seeds are pressed into oil plus a press cake conserving about two-thirds of the energy in the oil, which is then refined and submitted to esterification to produce methyl esters.

In this process oil reacts with light alcohols (methanol or ethanol) at atmospheric pressure and temperatures around 70°C. Cheap bases are used as catalysts. The purpose of the esterification is to lower oil viscosity. Ethanol is a preferred alcohol in this process because it has more favourable environmental features. This technology is well known and has been optimised specifically for biodiesel production since the early 1990s (Pandey, 2009), and perspectives for efficiency improvement are limited (Table 6.11).

6.7 Evaluation

Biofuel production

Development and performance of the biofuel industry are impressive. Ethanol production doubled in a period of 10 years, responding to an increased blending obligation. Production of biodiesel has shown a strong increase. By 2008 ethanol consumption in Brazil exceeded that of gasoline, which is probably unique in the world. In 2012 sugarcane ethanol represented some 18% of the country's transport energy consumption, as compared to 23% for gasoline and 49% for diesel ('Brazil', 2012).

Land use

Changes in land use since 2000 are presented in Table 6.12. The area for biofuel production increased by 4.9 million ha in 2010. One-third of this (1.8 million ha) is dedicated to cane cultivation; increased soybean cultivation

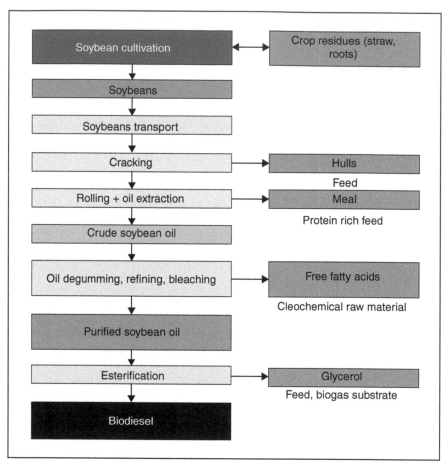

Figure 6.5 Conversion from soybean to biodiesel. Figure design by H. Croezen.

Table 6.11 Efficiency of soybean-to-biodiesel conversion in Brazil

	Unit	*BioGrace*	*Other Sources*
Conversion efficiency	litre/tonne	197	198–205
Conversion efficiency	GJ/GJ	0.33	
Biofuel yield	litre/ha	550	495–570
Biofuel yield	GJ/ha		16.3
Soy meal yield	tonne/ha		1.8
Glycerol yield	tonne/ha		0.9

Sources: Calculated from BioGrace (2012); Fediol (2012), http://www.fediol.eu; Greenfacts (2012).

Table 6.12 Changes in land use 2000–2010

	Unit	Impact on Land Availability	Share of Land Gain
Increase in biofuel production	million ha	−4.9	
Allocation to co-products	million ha	1.8	
Net biofuel land use	million ha	−3.1	
Area generated by increased MCI	million ha	4.9	29%
Change in agricultural area	million ha	12.0	71%
Gains arable land total	million ha	17.0	100%
Net land balance	million ha	13.8	

Sources: Calculated from Tables 6.1, 6.5, and 6.7 and FAOSTAT (2010–2013), http://faostat .fao.org.

accounted for 3.1 million ha. Two sources of additional land are identified. Expansion of agricultural area has added 12 million ha, almost four times the net biofuel area expansion. On top of this, increasing harvest frequency (i.e., MCI) generated an extra 4.9 million ha of crops. This alone has been more than sufficient to compensate for biofuel expansion. Expansion and intensification of agricultural area more than compensate for additional bio-fuel land use and, with it, land availability for food and feed markets has increased. Total land availability increased by nearly 14 million ha. Because intensification generated 1.6 times the amount of land that was 'lost' for biofuel production, the expansion of agricultural land (mostly arable land), strictly speaking, was not needed (even though it occurred).

Notwithstanding recent progress in reducing the loss of rainforests[2] and other sensitive biomes, Brazil still is facing important developmental challenges. Benefits of agricultural growth should be combined with the protection of the environment and sustainable development. Recent changes in deforestation legislation seem to be giving more room for farmers in the Amazon region to legally clear patches of forest.

Production efficiency

Metrics of the biofuel chains are rather favourable (see Table 6.13). Biofuel yield is high for sugarcane, while soybean has a relatively high biodiesel yield. Sugarcane does not generate any co-products. Crop and biofuel out-put per unit of nitrogen and phosphate is high to very high (soybean). Risks for nutrient leaching are low for sugarcane, a multiannual crop with an intensive rooting system. Application of agro-chemicals is low for sugarcane but high for soybean, as is the water requirement.

No-till practices and low nutrient applications for soybeans limit the risks for leaching. Sugarcane production systems, which combine from six

Table 6.13 Performance and impacts of sugarcane and soybean biofuel chains in Brazil

	Unit	Sugarcane	Soybean
Crop yield	tonne/ha	79.5	2.8
Biofuel yield	litre/ha	7,200	600
Biofuel yield	GJ/ha	152	16
Co-product yield	tonne/ha	–	1.8
Nitrogen productivity (crop)	kg dm/kg N	327	595
Water requirement (biofuel)	m³/GJ	37	145
Nutrient productivity (biofuel)	GJ/kg N	2.2	13.6
	GJ/kg P_2O_5	1.9	1.4
Risk of nutrient leaching	–	Low	Moderately high
Impact on soil organic matter	–	Positive	Moderately positive

to seven years of continuous cultivation with an inter-crop, have a positive impact on soil organic matter. The impact depends on the amount of soil carbon stored in the soil; soils high in clay and organic matter need higher amounts of residues to compensate for mineralisation. The impact will be larger if leaves are not burnt before harvest. Soybean impacts on soil carbon are expected to be slightly positive.

GHG emissions

Greenhouse gas emissions related to biofuel production are presented in Table 6.14. Emissions related to biofuel production have been estimated at about 12–24 g CO_2-eq. per MJ for sugarcane ethanol and 45–57 g CO_2-eq. per MJ for soybean biodiesel. This represents a reduction compared to the fossil fuel reference of about 70–80% for ethanol and 30–50% for biodiesel. Emissions associated to indirect land-use change (ILUC), estimated at 4–46 g CO_2-eq. per MJ ethanol, can seriously limit the reduction level. For soybeans, reported ILUC-related emissions are so high that total emissions may exceed those of fossil diesel.

It has been estimated that co-generation from sugarcane plants represent net savings in CO_2 emissions to an extent of 47 million tonnes, or 20%, of all fuel emission in Brazil (Macedo, 2008).

Economic and social impacts

Since the early twenty-first century Brazilians benefited from stable economic growth, relatively low inflation rates, and a substantial improvement of social well-being (World Bank, 2013). The favourable economic

Table 6.14 GHG emissions of Brazilian biofuel chains

	Unit	Cane	Soybean
Biofuel production	g CO_2-eq/MJ	12–24	45–57
Reduction[1]	%	70–80%	30–50%
ILUC	g CO_2-eq/MJ	4–46	16–55
Total	g CO_2-eq/MJ	16–70	61–112

Note: Biofuels for domestic use.
[1]ILUC-related emissions not included. Calculated with a fossil reference of 90 g CO_2-eq / MJ.

Sources: Calculated from BioGrace (2012); Laborde (2010); Darlington et al. (2013); Macedo and Seabra (2008); CARB (2012).

development was supported at least partly by the biofuel sector (Neves et al., 2010). Economic impacts of biofuel production are based on (1) replacement of fossil imports and some export earnings, (2) employment created in the industry, and (3) indirect income effects. These will be discussed here briefly.

The contribution of the cane sector to the national economy has been estimated at US$28 billion, equivalent to nearly 2% of the Brazilian GDP. This corresponds to the annual income of a country such as Uruguay (Neves et al., 2010). Contribution of the Proálcool programme investments, totalling US$11 billion since its inception, are estimated to have saved Brazil some US$27 billion in foreign oil imports (Elbersen et al., 2010).

Estimates of employment in the biofuel industry vary. According to Sawyer (2008), sugarcane cultivation and harvesting require approximately one million seasonal workers. Estimates by Krivonos and Olarreaga (2006) suggest that the combined sugar and alcohol sectors employed 765,000 people in 2002. Roughly half were working in cane cultivation, mostly in the central-south region. Increasing mechanisation has been reducing the need for low-skilled workers.

Dal Belo Leite et al. (2013) studied farming performance in Minas Gerais, which is a state characterised by a large number of agro-ecological zones and smallholdings, while an active bioenergy policy is implemented there to boost rural development and decrease the negative impacts of fossil fuel consumption. In the northern part of the state, a transition region from the Cerrado towards the semiarid north is formed under conditions of erratic rainfall and poor soil quality. In the northwest, Minas Gerais links to the Brazilian midwest, a major production region for soybean, maize, beans, and sorghum.

Using a farm typology, the authors studied opportunities for the production of feedstocks for the biodiesel industry. Part of the (national) biodiesel policy is oriented towards the production of alternative oil crops, including castor beans. Not all farmers can produce soybeans because soil quality and rainfall in some regions are not suited for it. The biodiesel industry is considered an attractive market, paying a bonus over prevailing soybean market prices to ensure feedstock availability.

Most family farms involved in biodiesel feedstock production appear to be soybean farmers. Non-soybean farmers, more common in number, could not link to the biodiesel industry. Farmers in semiarid regions have been encouraged to grow castor beans, a crop which is better able to withstand drought conditions while producing a higher share of oil. The authors conclude that the use of castor oil in the biodiesel industry is not very relevant.

Related to their position linked to biodiesel producers, soybean farmers generally showed better performance. They were facing fewer difficulties obtaining inputs and market access because biodiesel producers tended to provide seed and logistic support. Non-soybean farmers needed to focus more on production of food for their subsistence, which diverted them from cash-crop production. They were more frequently found in less-favoured areas encountering limited rainfall, poor soils, and so on. But also farmers that were better endowed showed limited willingness to engage in cash crops due to high transaction costs that involved more risks for food production and household economic well-being (Dal Belo Leite et al., 2013).

Wage levels in the sugarcane sector are relatively high (Walter et al., 2006), and cane expansion in the central-south is an important source of economic growth in the region (Sparovek et al., 2008). Working conditions in cane production, however, have been heavily criticised as being unhealthy, even leading to death by exhaustion (Sawyer, 2008). According to the ILO (2008), some 25,000–40,000 people are working in conditions akin to slavery.

The sugar industry seems, however, to be increasingly focusing on social issues, including the improvement of workers' quality of life, while also addressing sustainability issues referring to the rational use of land and water, the mitigating effects of mechanised harvesting, and the preservation of ecosystems (Neves et al., 2010).

Biomass availability

In comparison to 2000, the biomass use for biofuel production in Brazil increased by 187 million tonnes (Table 6.15), 178 million tonnes of sugarcane and 8 million tonnes of soybean. Only a limited amount of co-products (5 million tonnes of soybean meal) has been generated. The increase of arable crop output, doubling in 10 years, has been more than sufficient to compensate for the increased biofuel feedstock demand. Net availability of agricultural biomass (excluding biofuels) increased with 284 million tonnes, and food and feed production has not suffered. This is mainly the result of a large improvement of biomass productivity (BP).

Table 6.15 Changes in biomass availability 2000–2010

	Unit	Impact on Biomass Availability	Share of Biomass
Increase total biomass production	million tonne	466	
Biomass in biofuel production	million tonne	–187	100%
Generation of co-products	million tonne	5	2%
Net impact of biofuels	million tonne	–182	98%
Net change in biomass availability	million tonne	284	
Increase in BP	million tonne	478	

Note: Reduction of biomass availability for traditional food and feed use depicted by negative figures. Positive figures indicate a higher biomass availability for food and/or feed markets.

Source: Calculated from Table 6.5, Table 6.7 and FAOSTAT (2010–2013), http://faostat.fao.org.

6.8 Conclusion

Biofuel production in Brazil is adequate and sufficient to support ambitious domestic energy policies. Ethanol consumption – exceeding that of gasoline in 2008 – is among the highest in the world. Price fluctuations on oil and sugar markets, however, affect year-to-year conditions faced by ethanol producers and by consumers, which is managed by, *inter alia*, a high proportion of flex-fuel vehicles.

Impacts of the increasing production on water and soil resources are manageable. Fertiliser consumption is modest, especially for soybeans. Nutrient productivity is high to very high, again more for soybeans than for cane. The risks of nutrient leaching seem limited, but high applications of agrochemicals in soybean cultivation create a risk for groundwater quality.

Impact on soil carbon seems to be positive, but this highly depends on soil type, land management, and weather conditions. Large areas of grazing land have been converted to cane production, but as indicated in Figure 6.1, grazing land is abundant in Brazil. The export earnings from bioethanol have been significant.

On the whole, the economics of the Brazilian biofuel program appear positive. Support for the biofuel production chain has been modest, although it was a considerable investment for the country during the early days. The economic benefits, however, are quite impressive and more than compensate the direct investments that have been reported. Costs for agronomic and soil-related research have not been considered in this analysis.

Social implications of biofuel production are ambiguous. On the one hand, working conditions in the cane industry are extremely harsh. On the other hand, manual harvesting is being phased out. This is positive for workers' health but will limit employment conditions in the long run. A more balanced agricultural development in the Cerrado could help to improve economic and social conditions in that region.

Notes

1. For an explanation of C3 and C4 pathways, see Chapter 2.
2. Deforestation since 2000 amounts to 24 million ha, compared to 31 million ha in the 1990s (Table 6.1).

References

BioGrace. (2012). Retrieved 12 May 2012 from http://www.biograce.net/content/ghgcalculationtools/calculationtool

Brazil. (2012). *Wikipedia.* Retrieved 18 May 2012 from http://www.bioenergywiki.net/Brazil

Bressan, A., and Contini, E. (2007). *Brazil: A pioneer in biofuels.* Presentation given for the Farm Foundation, Washington, DC. Retrieved 8 January 2013 from http://www.farmfoundation.org/news/articlefiles/826-brazilpanelpres.pdf

Burrell A. (Ed.). (2010). *Impact of EU biofuel target on agricultural markets and land use.* Retrieved 12 January 2012 from http://publications.jrc.ec.europa.eu/repository/bitstream/111111111/15287/1/jrc58484.pdf

CARB. (2012). *Carbon intensity lookup table for gasoline and fuels that substitute for gasoline.* Retrieved 21 September 2013 from http://www.arb.ca.gov/fuels/lcfs/lu_tables_11282012.pdf

Climate of Brazil. (2013). *Wikipedia.* Retrieved 26 January 2013 from http://en.wikipedia.org/wiki/Climate_of_Brazil

C2ES (2012). *Ethanol.* Retrieved 12 February 2012 from http://www.c2es.org/technology/factsheet/Ethanol

Dal Belo Leite, J. G., Bijman, J., Giller, K. E., and Slingerland, M. A. (2013). Biodiesel policy for family farms in Brazil: One-size-fits-all? *Environmental Science Policy,* Vol 27, pp195–205.

Darlington, Th., Kahlbaum, D., O'Connor, D., and Mueller, S. (2013). *Land use change greenhouse gas emissions of European biofuel policies utilizing the Global Trade Analysis Project (GTAP) model.* Macatawa, USA: Air Improvement Resource.

De Almeida, E. F., Bomtempo, J. V., and De Souza, C. M. (2007). *The performance of Brazilian biofuels: An economic, environmental and social analysis.* Paris, France (OECD), Joint Transport Research Centre.

Do Amaral, W.A.N., Marinho, J. P., Tarasantchi, R., Beber, A., and Giuliani, E. (2008). Environmental sustainability of sugarcane in Brazil. In P. Zuurbier and J. Van de Vooren (Eds.), *Sugarcane ethanol: Contributions to climate change mitigation and the environment* (pp. 113–138). Wageningen, the Netherlands: Wageningen Academic Publishers.

El Bassam, N. (2010). *Handbook of bioenergy crops. A complete reference to species, development and applications.* London: Earthscan.

Elbersen, H.W., Bindraban, P.S., Blaauw, R., and Jongman, R. (2010). Biodiesel from Brazil. In H. Langeveld et al. (Eds.), *The biobased economy: Biofuels, materials and chemicals in the post-oil era* (pp. 283–301). London: Earthscan.

Embrapa. (2010). *Tecnologias de produção de soja – região central do Brasil 2011.* Sistemas de Produção 14, Embrapa Soja Londrina.

Ethanol fuel in Brazil. (2012). *Wikipedia.* Retrieved 18 May 2012 from http://en.wikipedia.org/wiki/Ethanol_fuel_in_Brazil

FAO. (2004). *Fertilizer use by crop in Brazil.* Land and Water Development Division. Rome, Italy: UN Food and Agricultural Organization.

FAO. (2008a). *Improving the efficiency of soil and fertilizer phosphorus use in agriculture.* Food and Agriculture Organization. Retrieved 30 September 2010 from ftp://ftp.fao.org/docrep/fao/010/a1595e/a1595e03.pdf

FAO. (2008b). *State of food and agriculture.* Rome, Italy: UN Food and Agricultural Organization

FAPRI-ISU. (2011). *World agriculture outlook.* Retrieved 18 May 2012 from http://www.fapri.iastate.edu/outlook/2011/

Gazzoni, D. (2007). *Overview of the Brazilian biodiesel industry: Present status and perspectives.* Presentation given at the workshop Biodiesel from Brazil: Technology and Sustainability, The Hague, Netherlands, 19 November.

Gazzoni, D. L., Henning, A. A., Garcia, S., and Lantmann, A. (2002). Guidelines for good agricultural practices: Soybean production. In *Guidelines for good agricultural practices, FAO/Embrapa.* Brasilia, Brasil, pp235–268.

Goedert, W. J. (1983). Management of the Cerrado soils of Brazil. *Journal of Soil Science,* Vol 34, pp405–428. DOI: 10.1111/j.1365–2389.1983.tb01045.x

Goldemberg, J., Coelho, S. T., and Guardabassi, P. (2008). The sustainability of ethanol production from sugarcane. *Energy Policy,* Vol 36, pp2086–2097.

Goldemberg, J., and Moreira, J. R. (1999). The alcohol program. *Energy Policy,* Vol 27, pp229–245.

Greenfacts. (2012). Retrieved 21 October 2012 from http://www.greenfacts.org/en/biofuels/figtableboxes/biofuel-yields-countries.htm

IFA. (2012). Sugarcane. Retrieved 20 October 2012 from www.fertilizer.org/ifa/content/.../8954/.../sugcane.pdf

IGBE. (2013). Retrieved 27 January 2013 from http://www.fas.usda.gov/pecad/highlights/2005/09/brazil_12sep2005/

ILO. (2008). *Climate change and employment case study no 1, Brazil and biofuels.* Retrieved 15 June 2009 from http://earthmind.net/labour/briefing/docs/ilo-2008-climate-change-case-study-01-brazil-and-biofuels.pdf

Krivonos, E., and Olarreaga, M. (2006). *Sugar prices, labor income, and poverty in Brazil.* Policy Research Working Paper Series 3874. Washington, DC: The World Bank.

Laborde, D. (2010). *Assessing the land use change consequences of European biofuel policies: Final report.* Washington, DC: International Food Policy Research Institute.

Macedo, I. C. (2008). *GHG emissions in the production and use of ethanol from sugarcane in Brazil.* Retrieved 8 November 2010 from http://www.iddri.org/Activites/Ateliers/081009_Conf-Ethanol_Presentation_Isaias_Macedo.pdf

Macedo, I. C., and Seabra, J. E. A. (2008). Mitigation of GHG emissions using sugarcane bioethanol. In P. Zuurbier and J. van de Vooren (Eds.), *Sugarcane ethanol* (pp. 95–112). Wageningen, the Netherlands: Wageningen Academic Publishers.

McVey, M., Baumel, P., and Wisner, B. (2000). *Brazilian soybeans: What is the potential?* Retrieved 27 January 2013 from https://www.extension.iastate.edu/agdm/articles/others/McVOct00.html

Meyer, D. E., and Cederberg, C. (2010). *Pesticide use and glyphosate-resistant weeds: A case study of Brazilian soybean production.* Gothenborg, Sweden: Swedish Institute for Food and Biotechnology (SIK).

Neves, M. F., Trombin, V. G., and Consoli, M. A. (2010). Measurement of sugar cane chain in Brazil. *International Food and Agribusiness Management Review,* Vol 13. Retrieved 26 January 2013 from http://ageconsearch.umn.edu/bitstream/93558/2/3.pdf

OECD. (2008). *Economic assessment of biofuel support policies.* Retrieved 19 June 2012 from http://www.oecd.org/dataoecd/54/10/40990370.pdf

OECD-FAO. (2012). *Agricultural outlook 2012–2021.* Paris, France: Organisation for Economic Co-operation and Development.

Pandey, A. (2009). *Handbook of plant-based biofuels.* Boca Raton, FL: CRC Press.

Sanchez, O. J., and Cardona, C. A. (2008). Trends in biotechnological production of fuel ethanol from different feedstocks. *Bioresource Technology,* Vol 99, pp5270–5295.

Sawyer, D. (2008). Climate change, biofuels and eco-social impacts in the Brazilian Amazon and Cerrado. *Philosophical Transactions B,* Vol 363, pp1747–1752.

Siebert, S., Portmann, F. T., and Döll, P. (2010). Global patterns of cropland intensity. *Remote Sensing,* Vol 2, pp1625–1643.

Sparovek, G., Barretto, A., Berndes, G., Martins, S., and Maule, R. (2008). Environmental, land-use and economic implications of Brazilian sugarcane expansion 1996–2006. *Mitigation and Adaptation Strategies for Global Change,* Vol 14, pp285–298.

Steenblik, R. (2007). *Subsidies: The distorted economics of biofuels.* The Global Subsidies Initiative. Geneva, Switzerland: International Institute for Sustainable Development.

Talamini, E., and Dewes, H. (2013). *The macro-environment for liquid biofuels in the Brazilian science, mass media and public policies.* Retrieved 26 January 2013 from http://cdn.intechopen.com/pdfs/17884/InTech-The_macro_environment_for_liquid_biofuels_in_the_brazilian_science_mass_media_and_public_policies.pdf

USITC. (2012). *Brazil: Competitive factors in Brazil affecting U.S. and Brazilian agricultural sales in selected third country market.* Washington, DC: United States International Trade Commission.

Valdes, C. (2011). Can Brazil meet the world's growing need for ethanol? *Amber Waves,* Vol 9, pp38–45.

Walter, A., Dolzan, P., and Piacente, E. (2006). *Biomass energy and bioenergy trade: Historic developments in Brazil and current opportunities.* Paris, France: IEA Task 40.

World Bank. (2013). *Brazil overview.* Retrieved 25 January 2013 from http://www.worldbank.org/en/country/brazil/overview

7 Biofuel production in the USA

J.W.A. Langeveld, P.M.F. Quist-Wessel and H. Croezen

7.1 Introduction

While Brazil paved the path for crop-based ethanol production during a period of three decades, the USA has effectively copied elements of existing biofuel policies and implemented them in their own agricultural production model. In the span of a few years, the USA has demonstrated that it is able to set policies in place and develop the necessary infrastructure to produce, trade, and consume ethanol nationwide without too many problems. This process has been driven by huge support. Still, it is a remarkable achievement. By doing so, the USA has profited from its scientific, economic, and industrial basis in agriculture and research and its capacity for production chain development. Even more remarkable is the fact that the USA has been able to surpass Brazil as the largest ethanol exporter, although it must be said that it was helped by adverse cropping conditions (droughts affecting crop yields) in Brazil and special conditions on the world market (high sugar prices tempted Brazil to convert its cane to sugar rather than into ethanol).

It is worthwhile to study the development of biofuel production in the USA in more detail. This chapter focuses on the USA's cultivation, production, and conversion of maize (i.e., corn) into ethanol and of soybean into biodiesel. It also dives into issues of sustainability and efficiency, including greenhouse gas (GHG) and (potential) land-use impacts of the recent and projected ethanol production in this country.

The chapter is organised as follows. First, it analyses land cover and land use in the USA (Section 7.2). Next, it discusses how the biofuel industry evolved in the USA (Section 7.3). Section 7.4 explains how domestic crops are involved in biofuel production, while Sections 7.5 and 7.6 elaborate on the dynamics of the production of two major biofuel feedstocks (corn and soybean) and their conversion to biofuels. This is followed by an evaluation of biofuel feedstock production (Section 7.7) and some conclusions (Section 7.8).

7.2 Land resources

The USA is a large industrial country with a strong agricultural sector. Growing population and economic development have led to enhanced urbanisation. Following decades of net fossil fuel imports, a policy change focussing on domestic bioenergy production was implemented, requiring additional research efforts on the production of liquid biofuels. The development and production of the necessary feedstocks, mainly corn starch, builds on a long tradition in agricultural research, including crop breeding, cultivation and conversion, and production-chain development that has been strong and very successful in the USA.

This research and chain development is supported by a rich land base of more than 900 million ha. Most of this (44%) is agricultural land; one-third is forest while 200 million ha is qualified as 'other' land (urban, industrial use, as well as wasteland). Figure 7.1 depicts a schematic overview of different types of land use.

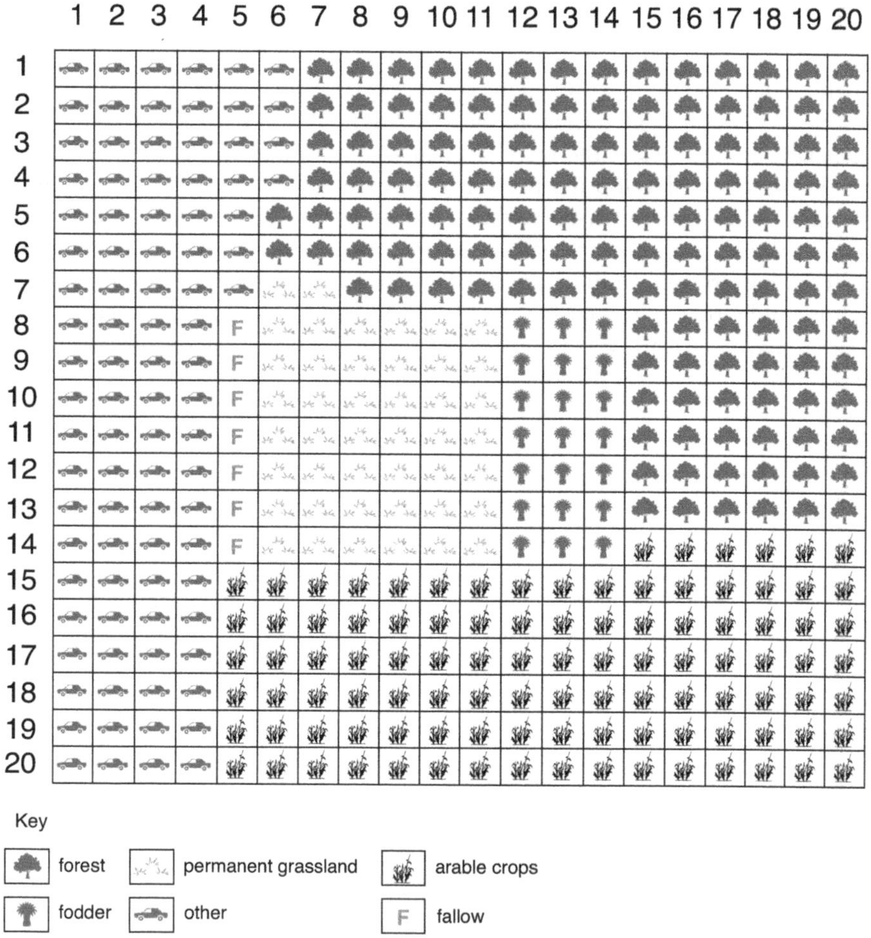

Key

forest   permanent grassland   arable crops

fodder   other   F fallow

Figure 7.1 Schematic overview of land use in the USA

No geographical representation. Each cell represents 0.25% of land area. Position of the categories was chosen randomly.

Source: Calculated from FAOSTAT (2010–2013), http://faostat.fao.org

Table 7.1 Land area and land use in the USA

	Unit	1980	1990	2000	2010
Forest area	million ha	No data	296	300	304
Other land	million ha	No data	193	202	200
Agricultural area	million ha	428	427	414	411
Permanent grassland	million ha	238	239	236	249
Agricultural tree crops	million ha	2	2	3	3
Arable area	million ha	189	186	175	160
Of which:					
arable crop area	million ha	115	103	104	102
fodder crop area	million ha	No data	No data	50	40
fallow	million ha	No data	No data	16[1]	15[2]
CRP	million ha			13[2]	13
Multiple Cropping Index (MCI)	–	0.74	0.72	0.77	0.84

[1] Figure for 2002.
[2] Figure for 2007.

Sources: Calculated from FAOSTAT (2012), faostat.fao.org; Siebert et al. (2010); ERD-USDA (2012); USDA (2011).

Details on land use are given by FAO (Table 7.1). Forest area is 304 million ha; 'other' land is 200 million ha. Agriculture covers 411 million ha. Nearly 60% of this is permanent grassland; 40% is classified as arable land. Some three million ha is covered with agricultural tree crops. Most arable land is devoted to arable crops, which include major food and feed crops such as corn, wheat, barley, soybean, and so on (102 million ha). One-third of arable land is used to produce fodder crops.

Fallow land was estimated at 15 million ha in 2007, which seems rather low. A special programme, the Conservation Rotation Program (CRP), was developed to limit soil erosion in the Midwest. It is a voluntary scheme that offers farmers compensation for not (ploughing and) cultivating part of their land and covered 13 million ha of arable land in 2010. Together, fallow and CRP cover over 20 million ha, or some 5% of agricultural land.

Soils in the USA are mostly high-activity clay soils with less productive clays in the east and volcanic and spodic soils locally in the west of the country. Dominant arable areas are located mostly on very productive clay soils. Climate conditions support good yields in many parts of the country, with water shortages being the most yield-limiting factor in the south and southwest, while relatively low temperatures limit the selection of crops in the north.

We calculated the Multiple Cropping Index (MCI) as an indication of the intensity of arable land use. MCI values in the USA are around 0.8, which means that in a given year, some 80% of all arable area actually is harvested. This is up from 0.72 in 1990 and 0.77 in 2000. Data from Table 7.1 further show an increase in forest and other land, while agricultural area shows a constant decline. Most losses are in arable land, which shrank with 29 million ha since 1980. Grassland area is increasing while fodder crops lost some 10 million ha. Fallow and CRP area did not change much over the past decade (data for earlier years are lacking).

These data suggest that arable land was converted to mostly forest and grassland. Farmers responded by reducing the share of fodder crops on arable land while agricultural tree crop area remained stable. The MCI has been increasing since 1988 but declined some in the 1990s. While nearly 30 million ha of arable land has been lost during the past three decades, the actual harvested area declined with no more than 13 million ha. Consequently, the 12 million ha loss of arable land was compensated by intensification. This means that a smaller area is cultivated more intensively. Since 2000 this has allowed farmers to harvest almost 11 million ha of arable crops without the need to expand the amount of land used.

7.3 Markets and policy conditions

USA biofuel development has mainly been defined by federal policies, which include minimum renewable fuel use and blending requirements as well as production tax credits, an import tariff for biofuels, loans, loan guarantees, and research grants (Schnepf and Yacobucci, 2012). The core of the policy is defined in the Renewable Fuel Standard (RFS), introduced under the Energy Policy Act of 2005. It was signed into law in 2007 when biofuel targets were defined as well as product quota and emission reduction targets.

In order to qualify as a renewable fuel, biofuels should realise GHG emission reduction of at least 20% compared to fossil fuels. Special catagories of biofuels (biodiesel, lignocellulosic ethanol, and ethanol produced in plants using biogas and reducing more than 50% of GHG reduction) may qualify as advanced fuels (Table 7.2). The Energy Independence and Security Act (EISA) further defines options for grandfathering and waivers and puts requirements on land-use change.

Specific fuel targets are presented in Table 7.3. Ethanol consumption, 34 billion litres in 2008, is to be gradually raised to 114 billion litres in 2022. A specific share is to come from advanced biofuels including lignocellulosic ethanol. Corn-based ethanol is not to exceed a maximum of 57 billion litres (half of the total 2020 consumption), but the share of lignocellulosic ethanol occasionally has been adjusted following limited availability, which remained less than projected volumes.[1] Total ethanol consumption is no more than 10% blending in gasoline. Under EISA, biodiesel is considered an advanced biofuel, and biodiesel targets are very modest.

Table 7.2 Federal mandates under the Energy Independence and Security Act (EISA)

	Corn Starch Ethanol	Advanced Biofuels		
		Cellulosic Biofuels	Biodiesel	Other
GHG reduction	>20% reduction[1]	>60% reduction[2]	>50% reduction[3]	>50% reduction[3]
Grandfathering	Older plants (construction started prior to 19 December 2007) can be exempt from environmental screens.			
Land-use change	Land must have been cleared prior to passage of law. No need for active use.			
Waiver options[4]	None. Most capacity met by grandfathered facilities.	Based on availability, not price. Permanent after 2016.	Based on price, maximum 120 days.	None

[1] Reducible to 10%.
[2] Reducible to 50%.
[3] Reducible to 40%.
[4] Waivers can be implemented if mandates lead to economic distress. Alternatively, totals can be scaled back pro-rata.
Source: Koplow (2009).

Table 7.3 Biofuel objectives in the USA

	Unit[1]	2006	2010	2012	2015	2020
Ethanol	billion litre	15	49	50	78	114
of which from corn starch	billion litre	–	45	50	57	57
Advanced biofuels	billion litre	–	3.6	7.6	21	57
of which cellulosic fuels	billion litre	–	2.5	3.3	11	40
of which advanced biodiesel	billion litre	–	2.5	4.8	4.8[2]	4.8[2]

[1] billion = thousand million
[2] Minimum – figure may be raised by EPA.

Sources: Adapted from Schnepf and Yacobucci (2012); C2ES (2009); Koplow (2009); Slingerland and van Geuns (2005).

A bill intended to phase out single-fuelled cars in favour of flex-fuel vehicles is under consideration. Several US car manufacturers sell flex-fuel cars. In 2006, six million vehicles were able to run on 85% ethanol blends. Several states further implemented additional biofuel policies. In early 2009, 37 states provided incentives promoting ethanol production and use while 9 states enacted their own renewable fuel standards. California has adopted a low carbon fuel standard (LCFS) that sets a goal of reducing the carbon intensity of passenger vehicle fuels (by a minimum of 10% by 2020). Other states are considering similar policies.

Target setting has led to a quick response. Biofuel production showed a spectacular development – corn ethanol production increased eightfold between 2000 and 2010 (Table 7.4). This is partly explained by the coincidence of cheap corn with high ethanol prices, especially in 2005 when margins for ethanol plants were extremely favourable. A third factor was the phasing out of methyl tertiary butyl ether (MTBE), a petroleum-based octane enhancer and oxygenate, as a gasoline additive in 2004 and 2005. MTBE fell out of favour because oil companies were held liable for potential damages to water supplies (Sanchez and Cardona, 2008; Babcock, 2011), a factor that is not related to biofuel policy.

US ethanol production was low before 2005 despite the existence of subsidies and low feedstock costs, as margins were kept low by low fossil energy prices. Profit margins showed a sharp increase in 2005, leading to a dramatic increase in investment and ethanol production. Most US ethanol

Table 7.4 Market development and economic conditions for biofuel production in the USA

	Unit[1]	Bioethanol		Biodiesel	
		2000	2010	2000	2010
Biofuel production	billion litre/ year	6.1	49.5	–	2.1
Biofuel consumption	billion litre/ year	6.2	48.8	–	1.8
Support investments	billion US$		7.0	–	
Subsidies, tax exemptions	US$/litre	$0.34	$0.13[2]	–	$0.27[2]
Import levies	US$/litre		$0.14	–	
Production cost	US$/litre	$0.50[3]	$0.36	–	$0.39–0.79[2]

[1] Billion = thousand million.
[2] Figure for 2007.
[3] Figure for 2004.

Sources: Calculated from FAPRI-ISU (2011); OECD-FAO (2007); 'Ethanol Fuel in Brazil' (2012); 'Ethanol Domestic Fuel Supply' (2012); APEC (2007).

plants were constructed starting in 2005, 2006, and 2007 in response to high profit margins, and new plants started to come online at the end of 2006. By 2009 construction of most plants was completed (Babcock, 2011). Ethanol production is sufficient to cover domestic consumption, and the USA became a net exporter in 2010. Further increases are expected. Biodiesel production rose to 2.1 billion litres in 2010. Consumption lagged behind, allowing export of surplus biodiesel production levels. FAPRI-ISU is projecting a further increase for biofuel production and consumption up to 2020.

Technological development and upscaling have helped to bring down ethanol production costs, which declined by one-third since 2000 but still were quite high at US$0.36 per litre in 2010. Further declines in production costs are expected. Domestic production is protected against cheap imports by an import levy of nearly 30% of production costs. Treaties may, however, give privileged access to a limited number of countries. Government support is estimated at US$7 billion in 2010. More support is the need to build infrastructure for the increasing number of ethanol plants.

7.4 Feedstock requirements

Grain maize (corn) and soybean are the most important biofuel crops in the USA. Ethanol production capacity has developed mainly in the corn-producing areas of the country; that is, the Midwest.

Corn for ethanol

Corn is the most important cereal crop in the world, and its importance is increasing. It is adapted to a wide range of climates but is mostly grown between latitudes 30° and 55°. It does well in most soils, although it does not thrive well in very heavy, dense clay and very sandy soils. Soils need to provide adequate drainage to allow for the maintenance of sufficient oxygen for good root growth and activity and enough water-holding capacity to provide adequate moisture throughout the growing season. Preferred pH is from 6.0 to 7.2 ('Maize/Corn', 2012).

Corn is mostly grown in the central area of the country (Figure 7.2). It is the most important arable crop in the USA, covering more than half of the cereal area. The main application is livestock feed. Data related to corn production are presented in Table 7.5. Annual production increased by 25% from 2000 to 2010, reaching a volume of more than 330 million tonnes. Half of this increase is attributed to yield increases. Corn area declined in the early 1980s but has been rising since 1983, reaching record highs after 2005. It has been projected to increase to 36 million ha in 2020. We calculated three-year average yields as a means to limit the impact of year-to-year weather impacts. Since 1980 yields have increased 50% to reach 10 tonne/ha. Average annual yield improvements were 60 kg per ha per year in the

☒ US Corn Belt

Figure 7.2 Main corn-producing regions in the USA

1980s, but they more than doubled after 1990. They are expected to remain at more than 100 kg per ha per year.

Corn use in the biofuel industry amounted to nearly 120 million tonnes in 2010 (up from 16 million in 2000), representing more than one-third of domestic production (Table 7.6). Area devoted to biofuel corn was raised with more than 9 million ha to reach 11.5 million ha. Corn use is projected

Table 7.5 Corn production in the USA

	Unit	*1980*	*1990*	*2000*	*2010*	*2020*
Harvested area	million ha	30	27	29	33	36
Production	million tonne	169	202	252	333	400
Yield (three-year average)	tonne/ha	6.6	7.2	8.6	10.3	11.1
Average annual increase[1]	kg/ha/year		60	140	130	108

[1] Figure for 1980–1990 presented in the column for 1990; figure for 1990–2000 presented in the column for 2000, and so on.

Sources: Calculated from FAOSTAT (2010–2013), faostat.fao.org; data for 2020 by FAPRI-ISU (2011).

Table 7.6 Corn use for ethanol production in the USA

	Unit	2000	2005	2010	2020
Use in biofuels	million tonne	16	36	119	133
Share of all corn	%	14	12	36	33
Biofuel feedstock harvested area	million ha	2.2	3.6	11.5	11.9

Sources: Calculated from OECD-FAO (2007); FAPRI-ISU (2011); FAOSTAT (2010–2012), faostat.fao.org; FAO (2008); Burrell (2010).

to show a further increase, but the area required for its cultivation is not expected to increase much.

Soybean for biodiesel

Soybean is an annual leguminous crop grown in tropical and (increasingly) temperate areas. The seeds contain 20% oil, which can be used for cooking and for the production of biodiesel. While soy oil is the dominant spice oil consumed in the USA, soybean meal is one of the most important animal feeds in the world because it provides an ideal mix of proteins and energy. Currently, 90% of the meal is used for feed. Strong increases in meat and livestock production in industrialised and emerging economies (China, Brazil, and India) have been based, for a large part, on the production of soy for feed purposes. The main production areas are Latin and North America.

Soybean production in the USA nearly doubled since 1980, supported by a 50% yield improvement (Table 7.7). Crop area increased to 31 million ha in 2010. Annual yield improvement amounts to some 30 kg per ha per year.

The use of soybean in the biodiesel industry increased from two million tonnes in 2005 to five million tonnes in 2010 (Table 7.8). It is projected

Table 7.7 Soybean production in the USA

	Unit	1980	1990	2000	2010
Harvested area	million ha	27	23	29	31
Production	million tonne	49	52	75	91
Yield (three-year average)	tonne/ha	2.0	2.3	2.6	2.9
Avg annual increase[1]	kg/ha/year	–	27	31	32

[1] Figure for 1980–1990 presented in the column for 1990; figure for 1990–2000 presented in the column for 2000, and so on.

Source: Calculated from FAOSTAT (2010-2013).

Table 7.8 Soybean use in biodiesel production in the USA

	Unit	2000	2005	2010	2020
Use in biofuels	million tonne	No data	2.0	5.0	6.1
Share of all soybean	%	No data	2.3	5.5	6.7
Biofuel feedstock harvested area	million ha	No data	0.7	1.7	1.9

Sources: Calculated from OECD-FAO (2007); FAPRI-ISU (2011); BioGrace (2012).

to increase by another million tonnes up to 2020. In 2010, 5.5% of all soybeans had been used to produce biofuels. That share is expected to rise to 6.7% in 2020. The main biodiesel production facilities are presented in Figure 7.3.

7.5 Crop cultivation

Common crop rotations in the USA include corn produced in a 1:2 rotation with a pulse crop, often soybean (when summers are long enough) or alfalfa (in cooler areas). A third crop (wheat) may be added to the rotation ('Maize', 2012). Crop rotations are applied in order to limit weed and insect damage and to reduce production costs.

Corn

Corn is an annual cereal crop of the C4 type, which means it is well suited to hot and dry conditions while its photosynthetic maximum is higher than that of temperate crops such as wheat (see Chapter 2 for background information on photosynthetic pathways in major biofuel feedstock types). It is well suited for the main agricultural areas of the USA, where it can realise yields that are among the highest in the world (Monfreda et al., 2008). In the upper Midwest, corn is sown between March and May and harvested between September and November in most years. A typical crop calendar is presented in Figure 7.4. There is little time after harvest, and catch crops normally are not sown.

Land preparation for corn is limited. In the Midwest farmers usually apply 'low-till' or 'no-till' techniques. Under low till, land preparation requires one or two runs, usually before planting or immediately after the harvest of the previous crop, after which the crops are planted and fertilised. No-till land is not treated, and seeds are sown immediately in crop residues. Both techniques reduce the loss of soil water and risk of erosion and subsequent loss of valuable soil organic matter ('Maize', 2012).

Corn yields are closely associated with nitrogen application levels. Sufficient nitrogen availability allows higher density, earlier sowing, and better

Activity	Jan	Feb	Mar	Apr	May	Jun	Jul	Aug	Sep	Oct	Nov	Dec
Sowing			X	X	X							
Growing period				X	X	X	X	X	X	X		
Harvesting								X	X	X		
Post-harvest											X	X

Figure 7.3 Cropping calendar for corn in the USA

weed control and moisture supply (FAO, 2006). Recommendations for nitrogen applications vary. As a rule, nitrogen application levels should be based on the potential yield, minus corrections for nutrient release from residues of previous crops or animal manure. In some regions, applications are further corrected for mineral nitrogen availability in the soil before sowing. Maize grown after a soybean crop needs 45 kg of nitrogen per ha less. Following alfalfa, this amounts to 100 kg of nitrogen per ha (IFA, 2012).

Corn receives some 145–170 kg of nitrogen per ha (Table 7.9). Farmers in the Northeast use slightly (5–10%) less mineral fertiliser than those in the eastern part of the so-called corn belt, as the former have a lower yield potential while more manure is available. Producers in the southeast face a lower yield potential but need relatively high amounts of fertilisers to compensate for weathered soils with lower inherent fertility (IFA, 2012). Low application figure by BioGrace is explained by the low yield levels applied in this model.

Adequate availability of phosphorus is essential because corn cannot readily take up large amounts of this nutrient that is essential for optimal growth and high yield. Applications should be varied according to available

Table 7.9 Corn input use in the USA

Parameter (unit)	Per ha	Source
Nutrient application (kg/ha)		
N	145–170	IFA (2012)
	52	BioGrace (2012)
P_2O_5	55–85	IFA (2012)
K_2O	55–80	IFA (2012)
Application of agro-chemicals	2.4	BioGrace (2012)
(kg active ingredient/ha)	2.4	James A. Baker III Institute for Public Policy (2010)
Diesel (litre/ha)	100	BioGrace (2012)
Required rainfall (mm)	500–800	FAO (2012)

soil phosphorus. In the corn belt, applications vary between 55 and 85 kg of P_2O_5 per ha (Table 7.9). Potassium is taken up in large quantities, but only a small proportion is removed with the grain. It is important to ensure that overall supply of potassium is sufficient to support high yields, especially when high nitrogen applications are given and high yields are expected. Again, it is advocated to vary application on the basis of soil analysis and yield potential (FAO, 2006). In practice, potassium fertiliser applications vary between 55 and 80 kg of K_2O per ha (IFA, 2012).

A majority of the corn has been genetically engineered to be resistant to herbicides and a number of insects.[2] The average application level of agro-chemicals amounts to 2.4 kg of active ingredient per ha. Diesel consumption reported by BioGrace is some 100 litre per ha.

Corn is an efficient user of water in terms of total dry matter production, and among cereals it is potentially the highest-yielding grain crop. For maximum production, a medium maturity grain crop requires between 500 and 800 mm of water depending on climate. To this, water losses during conveyance and application must be added. Corn appears relatively tolerant to water deficits during the vegetative and ripening periods. The greatest decrease in grain yields is caused by water deficits during the flowering period, due mainly to a reduction in grain number per cob. This effect is less pronounced when the plant suffered water deficits in the preceding vegetative period. Water deficits during the yield formation period may lead to reduced yield due to a reduction in grain size. Water deficit during the ripening period has little effect on grain yield (FAO, 2012).

Soybean

Soybean is a tropical annual C3 pulse crop, which is the most common source of proteins used in human food and animal feed.[3] It is grown in large parts of the Midwest where it reaches high to very high yields. The cropping calendar is similar to that of corn. A typical crop calendar is presented in Figure 7.5.

Soybeans require well-drained soils with adequate water retention capacity and fertility. Ideally, soils should be of medium texture (clay content from 30% to 35%) or heavy but well-drained soils. Soil depth should be more

Activity	Jan	Feb	Mar	Apr	May	Jun	Jul	Aug	Sep	Oct	Nov	Dec
Sowing			X	X	X							
Growing period				X	X	X	X	X	X	X		
Harvesting									X	X	X	
Post-harvest											X	X

Figure 7.4 Cropping calendar for soybean cultivation in the USA

Table 7.10 Input use in soybean cultivation in the USA

Parameter (unit)	Per ha	Source
Nutrient application (kg)		
N	0	IFA (2012)
	3	Pradhan et al. (2011)
P_2O_5	32–42	Medium phosphate soil level. IFA (2012)
	28	Pradhan et al. (2011)
K_2O	55–105	Medium potassium soil level. IFA (2012)
	27	Pradhan et al. (2011)
Application of agro-chemicals (kg of active ingredient/ha)	1.6	Pradhan et al. (2011)
Diesel (litre/ha)	33	Pradhan et al. (2011)
Gasoline (litre/ha)	13	Pradhan et al. (2011)
Required rainfall (mm)	500–750	El Bassam (2010)

than 1 metre. Soils with low (<15%) clay content should be avoided, and the crop should not be sown in soils with slopes greater than 8% when using conventional cropping systems (or more than 15% when using the no-till systems). Soils with major physical problems, like the presence of stones or high salt concentrations, should also be avoided (Gazzoni et al., 2002).

According to IFA (2012), there is no need for nitrogenous fertilisers, but in practice some seem to be applied (Pradhan et al., 2011). An overview of input use is presented in Table 7.10. Application rates of phosphate and potassium fertilisers depend on the nutrient status of the soil. In practice, some 30 kg of P_2O_5 and K_2O is given, which indicates that soils in the USA generally have adequate nutrient levels. Application of agro-chemicals is just more than 1.5 kg of active ingredient per ha – a low amount. Consumption of fossil fuels (46 litre/ha) is modest.

7.6 Conversion to biofuels

Corn to ethanol

Corn-to-ethanol conversion routes are presented in Figure 7.6. Corn grain can be processed through 'dry' or 'wet' milling and consequent fermentation and distilling of grains. Wet milling starts with water-soaking the grain and then adding sulphur dioxide to soften the kernels and loosen the hulls, after which it is ground. It uses well-known technologies and allows separation of starch, cellulose, oil, and proteins. Dry milling grinds whole grains (including germ and bran), mixing the flour with water to be treated with liquefying enzymes and,

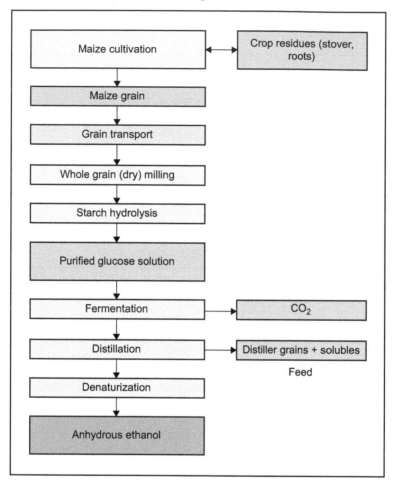

Figure 7.5 Conversion from corn to ethanol. Figure design by H. Croezen.

further, cooking the mash to break down the starch. This hydrolysis step can be eliminated by simultaneously adding saccharifying enzymes and fermenting yeast to the fermenter (simultaneous saccharification and fermentation).

After fermentation, the mash is sent through a distillation system, followed by concentration, purification, and dehydration of the alcohol. The residue mash ('stillage') is separated into a solid (wet grains) and liquid (syrup) phase that can be combined and dried to produce dried distillers grains with solubles (DDGS) to be used as cattle feed. Its nutritional characteristics and high vegetable fibre content make DDGS less suited for other animal species, and extension to the more lucrative poultry and pig-feed markets still is a challenge.

Table 7.11 Efficiency of corn-to-ethanol conversion in the USA

	Unit	Biograce	Other Sources	
			2010	2020
Conversion efficiency	litre/tonne	381	377	409
Conversion efficiency	GJ/GJ	0.52		
Biofuel yield	litre/ha		3,730	
Biofuel yield	GJ/ha		80.3	
DDGS yield	tonne/ha		1.6	

Sources: Calculated from BioGrace (2012); Elsayed et al. (2003); FAO (2008); Burrell (2010).

Ethanol contains only half of the energy of the corn feedstock. Most of the remainder is recovered in the co-products (Table 7.11). Corn-to-ethanol conversion in 2010 was 377 litre per tonne, which under average conditions in the USA yields some 3,700 litres of ethanol per ha. Conversion efficiency has been projected to improve to exceed 400 litres per tonne in 2020 (Burrell, 2010). Theoretical corn ethanol yield maximum is 520 litres per tonne. DDGS is the key co-product in the dry-grind ethanol process, which can bring additional revenues when sold as animal feed. Corn oil, corn gluten meal, and corn gluten feed also are important co-products in the wet-mill process (Drapcho et al., 2008).

Soybean to biodiesel

Conversion for biodiesel production from soybean is depicted in Figure 7.7. The process consists of three steps. First, seeds are pressed into oil plus a press cake, conserving about two-thirds of the energy in the oil, which then is refined and submitted to esterification to produce methyl esters. In this process, the oil reacts with light alcohols (typically methanol or ethanol) at atmospheric pressure and temperatures around 70°C. Cheap bases are used as catalysts. The purpose of the esterification is to lower the viscosity of the oil. Ethanol is a preferred alcohol in this process because it has more favourable environmental features. This technology is well known and has been optimised specifically for biodiesel production since the early 1990s (Pandey, 2009), and perspectives for efficiency improvement are limited.

Soybean biodiesel yields are low in comparison to rapeseed, for example, which is explained by its relatively high protein content (Drapcho et al., 2008). Conversion efficiency is some 200 litres of biodiesel per tonne of seeds (Table 7.12). One-third of the energy in the seeds is recovered in biodiesel end product. Co-products are press cake and glycerol. Average

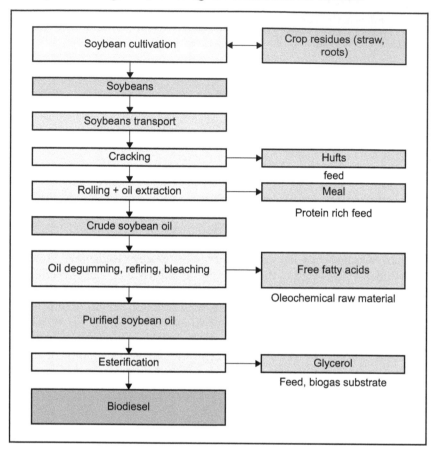

Figure 7.6 Conversion from soybean to biodiesel. Figure design by H. Croezen.

biodiesel yield for the USA is about 570 litres per ha. Biodiesel conversion efficiency generally is not expected to improve much in the near future.

7.7 Evaluation

Biofuel production

Since the introduction of the biofuel policy, industry in the USA has developed extremely fast. The country now is an exporter of ethanol (competing with Brazil) and is close to self-sufficient in biodiesel. After a slow start, production capacity for lignocellulosic biofuel production now is developing. High production costs remain a source of concern, but further reductions are expected.

Table 7.12 Efficiency of soybean-to-biodiesel conversion in the USA

	Unit	Biograce[1]	Other Sources
Conversion efficiency	litre/tonne	197	198–205
Conversion efficiency	GJ/GJ	0.33	
Biofuel yield	litre/ha	550	495–570
Biofuel yield	GJ/ha		16.3
Soy meal yield	tonne/ha		6.5
Glycerol yield	tonne/ha		3.2

[1] Figures refer to soybean production in Brazil.

Sources: Calculated from BioGrace (2012); Fediol (2013); 'Table 2: Biofuel Yields' (2012).

The development of a biofuel industry has been supported by high increases in feedstock availability. This holds especially for corn production, which showed a dramatic increase. This can be partly explained by yield improvement, although yield increases have in fact remained in line with historic trends. In 2010 more than one-third of corn was applied in the biofuel industry. Almost 120 million tonnes of corn has been used, occupying more than 11 million ha while replacing less productive crops.

Soybean is grown in similar rotations as corn, generally making use of the same soils and undergoing similar conditions. It requires hardly any nitrogen and only low applications of other nutrients. Soybean meal is an important output for the livestock industry. Its development during the past few years has been less spectacular, but still more than 1.5 million ha of the crop is devoted in biofuel production.

Land use

Impacts of biofuel expansion on land use are presented in Table 7.13. From a total of 11 million ha used to produce biofuel crops, nearly six million ha is associated to the co-product. Just more than five million ha is no longer available for food production. Loss of agriculture area in the USA amounted to 3.5 million ha. Arable land partly was converted to forest; the remainder was allocated to urban, recreational, or industrial use, or was left idle.

Harvesting frequency (MCI) has increased. Raising MCI has facilitated an additional harvested area of more than 10 million ha of arable crops per year. Because of this, net land availability increased with 2.3 million ha.

Production efficiency

Corn is a moderately high-yielding ethanol crop, which generates a high co-product output (Table 7.14). It receives rather high fertiliser applications, especially the nitrogen that is crucial for good yields. Nitrogenous

Table 7.13 Changes in land availability for food and feed production in the USA 2000–2010

	Unit	Impact on Land Availability	Share of Land Loss
Increase in biofuel production	million ha	−11.0	
Allocation to co-products	million ha	5.9	
Net biofuel land use	million ha	−5.1	59%
Loss of agriculturale area	million ha	−3.5	41%
Loss of arable land total	million ha	−8.6	100%
Area generated by increased MCI	million ha	10.9	
Net land balance	million ha	2.3	

Source: Calculated from Tables 7.1, 7.6, and 7.8, FAOSTAT (2010–2013), faostat.fao.org.

Table 7.14 Performance and impacts of corn and soybean biofuel chains in the USA

	Unit	Corn	Soybean
Crop yield	tonne/ha	9.9	2.8
Biofuel yield	litre/ha	3,800	550
Biofuel yield	GJ/ha	80	18
Co-product yield	tonne/ha	4.2	1.8
Nitrogen productivity (crop)	kg dm/kg N	50	238
Nutrient productivity (biofuel)	GJ/kg N	0.9	5.4
	GJ/kg P_2O_5	1.7	1.4
Water requirement (biofuel)	m^3/GJ	73	145
Risk of nutrient leaching	–	High	Moderately high
Impact on soil organic matter	–	Positive	Negative

compounds (especially nitrate) form a risk for nutrient leaching. Biofuel yield per unit of nitrogen fertiliser is rather low, but response to phosphate applications is good.

Impact on soil organic matter depends on a range of factors. No-till, no stover removal, and low soil erosion all help to improve the soil carbon. Generally, soils with high clay contents require larger residue contributions to maintain high organic matter stocks. Under prevailing conditions of no-till, high yields, and no removal of stover, corn production on low clay soils will have a positive impact on soil organic matter (carbon sequestration).

Being an oil crop, soybean has a low crop and biofuel yield, but it generates relatively high shares of meal. As a pulse crop, it contributes nitrogen to the soil. Output per kg of nutrient is good (phosphorus) to very good (nitrogen), but corn provides more bioenergy per unit of phosphorus due to its high yields. Contribution to soil organic matter is lower and insufficient to compensate for mineralisation (net negative soil carbon balance).

The increase of biofuel feedstock cultivation may have implications for nutrient use and emissions. Fertilization levels of corn are higher, and this crop offers higher risks for losses of nutrients than some other crops it may be replacing. FAO data suggest that corn expansion has been coinciding with the replacement of other cereals (mainly barley) as well as fodder crops (not of permanent grasslands). The impact depends on the way fertilization of corn and soybean differs from that of barley and fodder crops. As corn requires high fertiliser applications, this crop will probably have a negative impact on water quality, but the exact effect will depend on local conditions (soil quality, including clay content, yield potential, and slope). Differences in the application levels of agro-chemicals will be modest as will be – under prevailing conditions – the impact on soil organic matter.

GHG emissions

GHG emissions related to biofuel production are presented in Table 7.15. Emissions from corn ethanol production vary from 32 to 74 g CO_2-eq/MJ. Emission reduction ranges between 23% and 64% (not including indirect land-use change [ILUC]). Reductions are limited if ILUC emission estimates are included (12–104 g CO_2-eq/MJ). The highest ILUC estimates have been reported by Searchinger et al. (2008), suggesting ILUC-related emissions of more than 100 g CO_2-eq/MJ, thus making ILUC emissions alone higher than those of the fossil fuels that are to be replaced. More recent reports suggest ILUC emissions of 10–12 g CO_2-eq/MJ.

Emissions related to the production of soybean biodiesel vary between 21 and 57 g CO_2-eq/MJ, with GHG reduction levels of 36–76%. ILUC emissions are 16–56 g CO_2-eq/MJ. High ILUC emission estimates that were reported (CARB, 2012; Laborde, 2010) have been challenged (e.g.,

Table 7.15 Suggested GHG emissions occurring during biofuel production

	Unit	Corn	Soybean
Biofuel production	g CO_2-eq/MJ	32–74	21–57
Reduction	%	23–64%	36–76%
ILUC	g CO_2-eq/MJ	12–104	16–56
Total	g CO_2-eq/MJ	43–178	48–112

Note: Biofuels for domestic use.
[1]ILUC- related emissions not included. Calculated with a fossil reference of 90 g CO_2-eq/MJ.

Sources: Calculated from BioGrace (2012); Laborde (2010); CARB (2012); Darlington et al. (2013).

Darlington et al., 2013). It probably remains too early to provide final estimates on emissions related to land-use change.

Economic and social impacts

Estimates of the economic impact of biofuel production are not easy to make. According to some sources, the ethanol industry created 150,000 jobs in 2005, which is before most of the plants were constructed and taken into use. This would have boosted household income by US$5.7 billion and tax revenues with some $3.5 billion at the local, state, and federal levels ('Energy Policy of the United States', 2013). A more recent estimate reports a contribution of ethanol production to the state economy in North Dakota of US$544 million (Voegele, 2013).

Biofuel exports and replacement of fossil imports positively affect the national economy. Gan et al. (in press) calculated farm income of stover ethanol production for Palo Alto County in Iowa at US$0.19 per litre while additional regional incomes amounted to US$0.42 per litre for the regional economy (including incomes from plant construction and operation and management jobs) plus a US$0.13 security premium for the national economy. Clearly, this can add up to huge amounts if lignocellulosic ethanol progresses the way it has been mandated.

These figures suggest that public support for biofuels has considerable positive economic impacts. They are in line with ERS data suggesting a US$0.29 domestic business activity in 2010 for every dollar of US agricultural exports. Total employment related to agricultural exports, including biofuels, amounted to 900,000 (USDA ERS, 2012).

Domestic biofuel production in the USA (as well as other biofuel production elsewhere) has, however, also contributed to price increases of main food commodities. While the exact contribution of biofuels is subject to debate, no doubt biofuel policies have played a role in the price spike that was observed toward the end of 2007 and in 2008. After this, prices declined again, but they remained at levels above those of the previous years.

According to one estimate, 2011 corn prices would have been 17% lower if the USA had not subsidised or supported domestic biofuel production. This effect will be felt by poor consumers in corn-importing countries, and institutions such as the World Bank suggest that biofuel mandates should be adjusted when food is short or prices are inordinately high. The largest impact of subsidies occurred in the 2007 when maize prices would have been 7.1% lower than they actually were. This is a modest impact because the average maize price in 2007 was more than US$2.00 per bushel higher than the average price in either 2004 or 2005. This implies that ethanol subsidies have not been the major driver of higher commodity prices. The effects of US ethanol subsidies on the prices of wheat, rice, and soybeans were even smaller, with a 2.8% price impact on soybeans in 2008 being the largest impact on these other crops. The impact of US ethanol policies through higher feed costs on consumer prices of eggs, beef, pork, and broilers was even smaller. These results indicate that US ethanol subsidies during this period had little impact on consumer prices and quite modest impacts on crop prices (Babcock, 2011).

If US ethanol production had somehow not been allowed to expand beyond 2004 levels, then maize prices in 2009 would have been about 21% lower than they actually were. Wheat and soybean prices in 2009 would have been about 9% and 5% lower, respectively. These results indicate that ethanol indeed contributed significantly to higher crop prices and modestly to higher food prices. But this does not imply that ethanol subsidies had the same effect. Higher crude oil prices would have increased the demand for biofuels and would have created strong market-driven investment incentives that would have resulted in a large expansion of the US ethanol industry even without the subsidies (Babcock, 2011).

It is stressed here that strong domestic demand growth due to expansion of its flex-fuel vehicle fleet has driven Brazilian ethanol prices to levels higher than prices in the USA. This means that removal of the US import tariff on ethanol would have no impact on trade flows. Ethanol exports to Brazil would be higher if US ethanol subsidies would be eliminated because the difference between Brazilian and US domestic prices would be even greater (Babcock, 2011).

The situation is different for biodiesel production from soybean oil, which under market conditions would not be economic without a mandate. The cost of producing biodiesel from soybean oil would be prohibitive. Soybean oil prices would fall by an average of 16% if the mandate and tax credit were eliminated. But soybean prices would then drop by an average of 3.2% because higher soybean meal prices would offset much of the effects of lower soybean oil prices (Babcock, 2011).

Different studies have estimated the impacts for food-importing countries and local poor consumers who spend more on food than on anything else. Wise (2012) estimated costs of corn price increases for corn-importing countries at some US$2 billion per year, with developing countries absorbing more than half of those costs.

Changes in biomass availability

Since the introduction of the biofuels policy, agricultural production in the USA has changed. Annual biomass use for biofuels in 2010 amounted to 119 million tonnes of corn and 5 million tonnes of soybean. Arable land has been lost, while the area of grassland and urban land has increased. These changes have been steered by many driving forces including – although not exclusively – biofuel policies. Land use and agricultural production are complex processes, and their dynamics are not easy to understand. Many factors play a role, and the number of actors involved is very high.

An analysis of biomass output per ha or arable land shows that arable crop production was raised from 2.6 to 3.2 tonne per ha in a period of only 10 years (calculated from FAOSTAT). In the 1990s average annual increase

Table 7.16 Changes in biomass availability in the USA since 2000

	Unit	Impact on Biomass Availability	Share of Biomass
Increase total biomass production	million tonne	48	
Biomass used in biofuel production	million tonne	−108	100%
Generation of co-products	million tonne	58	54%
Net impact of biofuels	million tonne	−50	46%
Net changes in biomass availability	million tonne	−2	
Increase in BP	million tonne	138	

Notes: Reduction of biomass availability for traditional food/feed use depicted by negative figures. Positive figures indicate a higher biomass availability for food and/or feed markets.

Source: Calculated from Table 7.6, Table 7.8 and FAOSTAT (2010–2013), http://faostat.fao.org.

of biomass productivity (BP) amounted to nearly 50 kg per ha per year. This has been raised to 70 kg per ha per year since 2000. Changes in BP are the end result of a number of process changes, which may include increasing input use, improving crop management, and changing weather conditions, but interpretation of these changes is not easy.

Table 7.16 gives an overview of changes in arable crop production and its availability for the food and feed markets since 2000. Biofuel production increase utilised 108 million tonnes of corn and soy, half of which was converted to co-products. As total arable production increased with 48 million tonnes, net biomass for food and feed availability has slightly declined. These findings seem to contradict results reported by Oladosu et al. (2011), who found that only 20% of the corn used in biofuel production was obtained from additional domestic corn production in the USA. The remainder was mostly from reallocation of existing corn use.

7.8 Conclusion

The USA has developed an effective biofuel policy and industry within a brief period of time. Biofuel production is largely based on domestic feedstocks, mainly corn and soybean. Half of biodiesel is made from other feedstocks, which may be partly imported. Ethanol production exceeded domestic consumption in 2012, making the USA the largest producer in the world as well as a net exporter.

These are no small achievements. They result from a balanced but also more-or-less flexible policy oriented at developing and supporting a domestic biofuel industry that is based on arable crop production under existing commercial frameworks and involving old and new industrial stakeholders. Farmers and plant owners (which sometimes are owned by farmers) responded well to incentives that were provided, raising corn output by 26% and soybean production by 21% in just 10 years and generating production capacity for billions of litres of biofuels.

Corn area increased by 12% and soy area by 6%, which is remarkable if one considers the fact that the amount of arable land available to farmers declined by nearly 13 million ha from 2000 to 2010. Corn and soybean yields continued to increase, but yield improvement did not accelerate in comparison to increases realised in the 1990s. Almost half of the 106 million tonnes of corn and soybean used for biofuels could be compensated by production of co-products (mainly DDGS and soy meal). While there are limitations in the applicability of DDGS as an animal feed (partly replacing either corn or soybean), much research currently is being done about improving DDGS digestibility. Soybean meal, of course, has well-known feed characteristics.

Yield increases and crop replacement thus seem to have modified the impact of biofuel crop expansion on land use in the USA. While biofuel expansion is responsible for 80% of the reduction in biomass availability for the food and feed market (even when co-products are considered), the situation on the *land* market is different. The amount of land used for biofuel feedstocks is almost equal in size to the amount of agricultural (arable) land that has been lost since 2000 (11.0 vs. 10.9 million ha). This means that apart from biofuel crop expansion, there is another driving force, which has reduced land availability for food and feed production. This is related to urbanisation and expansion of industrial, infrastructural, and leisure land use, although also other changes may have occurred (increase of forest area).

Another contributor to the enhanced availability of feedstocks has been a structural change in the land-use system. Throughout the past decade, farmers harvested an extra 14 million tonnes of arable crops from the same amount of arable land. This can partly be attributed to the replacement of fodder crops. It is also caused by yield improvement and replacement of less productive crops by higher-yielding alternatives. There is evidence that corn expansion has occurred at the expense of barley, even though the growing areas of both crops only partly coincide. Since 2000 one million ha of barley has been replaced. Because corn yield is 2.5 times higher this alone has generated an extra 6.5 million tonnes of biomass, which is equivalent to 6% of the total biofuel feedstock use in 2010. (Note that the difference is larger if one considers the fact that corn stover yield is also 2.5 times higher than that of barley.)

Enhanced biofuel production has positive economic impacts. Extra jobs and economic growth have been generated while ethanol exports and replacement of fossil fuel imports have a positive impact on the national economy. Arable farmers in the USA profited from higher crop prices, but livestock producers literally paid a price for this development.

The impact of biofuel area expansion on the environment, however, should not be neglected. Corn consumes relatively high levels of fertilisers, and it is a known source of high nutrient emissions. The impact of changes in crop rotations on the quality of soil and ground and surface water should be studied and considered. In terms of energy production per unit of nutrient use, corn is rather efficient (be it that soybean – a leguminous crop – is more efficient in nitrogen use).

Arable farmers and regional and national economies have gained from the emerged biofuel industry in the USA, but this has come at a price. Investment subsidies have been taken from the federal budget, which indirectly affected consumers and companies in the USA. Both have also been able to profit from lower fuel prices at the pump due to the blending with domestic biofuels. This advantage did not necessarily apply to inhabitants of fossil fuel–exporting or US corn-importing nations.

While it is claimed that poor consumers in developing countries who depend on import of corn for food or livestock may be affected by enhanced biofuel production, the exact effect will depend on the way corn (soybean) feedstock use for biofuels has affected price and availability of these crops for food export markets. A preliminary analysis of cereal and oilseed trade balances suggests that volumes did not change much. This is in line with increases in BP that were calculated previously, as these actually exceed net biomass demand for biofuel production. Still, it remains difficult to get a clear view of biomass production and use dynamics and their relationship to changes in biofuel development as compared to other changes; for example, land-use changes caused by urbanisation, tourism, or nature development (cf. Dale and Kline, 2013).

Notes

1. In December 2011 the EPA revised an initially mandated 500 million gallons for 2012 down to 8.7 million gallons. It remains uncertain if this will be met ('Waste to Biofuels Market Analysis 2013', 2012).
2. From http://cropwatch.unl.edu/web/bioenergy/corn.
3. For an explanation of C3 and C4 pathways, see Chapter 2.

References

APEC. (2007). *United States biofuels activities*. Retrieved 15 January 2013 from http://www.biofuels.apec.org/me_united_states.html

Babcock, B. A. (Ed.). (2011). *The impact of US biofuel policies on agricultural price levels and volatility*. ICTSD Programme on Agricultural Trade and Sustainable Development, Issue Paper No. 35. Geneva, Switzerland: ICTSD International Centre for Trade and Sustainable Development.

BioGrace. (2012). Retrieved 12 May 2012 from http://www.biograce.net/content/ghgcalculationtools/calculationtool

Burrell, A. (Ed.). (2010). *Impact of EU biofuel target on agricultural markets and land use.* Retrieved 12 January 2012 from http://publications.jrc.ec.europa.eu/repository/bitstream/111111111/15287/1/jrc58484.pdf

CARB (California Air Resources Board). (2012). Carbon intensity lookup table for gasoline and fuels that substitute for gasoline. Retrieved 21 September 2013 from http://www.arb.ca.gov/fuels/lcfs/lu_tables_11282012.pdf

C2ES. (2009). *Ethanol.* Retrieved 12 February 2012 from http://www.c2es.org/technology/factsheet/Ethanol

Dale, V.H., and Kline, K.L. (2013). Issues in using landscape indicators to assess land changes. *Ecological Indicators.* Retrieved from http://dx.doi.org/10.1016/j.ecolind.2012.10.007

Darlington, Th., Kahlbaum, D., O'Connor, D., and Mueller, S. (2013). *Land use change greenhouse gas emissions of European biofuel policies utilizing the Global Trade Analysis Project (GTAP) model.* Macatawa, USA: Air Improvement Resource.

Drapcho, C.M., Nhuan, N.P., and Walker, T.H. (2008). *Biofuels engineering process technology.* New York: McGraw-Hill.

El Bassam, N. (2010). Handbook of bioenergy crops. A complete reference to species, development and applications. London: Earthscan.

Elsayed, M.A., Matthews, R., and Mortimer, N.D. (2003). *Carbon and energy balances for a range of biofuels options.* Sheffield Hallam University.

Energy policy of the United States. (2013). *Wikipedia.* Retrieved 15 January 2013 from http://en.wikipedia.org/wiki/Energy_policy_of_the_United_States

ERD-USDA. (2012). Retrieved 12 January 2012 from http://www.ers.usda.gov/Briefing/LandUse/agchangechapter.htm

Ethanol domestic fuel supply or environmental boondoggle? (2012). *Scientific American.* Retrieved 11 June 2012 from http://www.scientificamerican.com/article.cfm?id = ethanol-domestic-fuel-supply-or-environmental-boondoggle

Ethanol fuel in Brazil. (2012). *Wikipedia.* Retrieved 19 May 2012 from http://en.wikipedia.org/wiki/Ethanol_fuel_in_Brazil

FAO. (2006). *Nutrient management guidelines for some major field crops.* Rome, Italy: Food and Agricultural Organization of the United Nations.

FAO. (2008). *The state of food and agriculture. Biofuels: Prospects, risks and opportunities.* Rome, Italy: Food and Agricultural Organization of the United Nations.

FAO. (2012). *Crop water information: Maize.* Retrieved 30 October 2012 from http://www.fao.org/nr/water/cropinfo_maize.html

FAPRI-ISU. (2011). *World agriculture outlook.* Retrieved 18 May 2012 from http://www.fapri.iastate.edu/outlook/2011/

Fediol. (2013). Retrieved 12 January 2013 from http://www.fediol.eu

Gan, J., Langeveld, J.W.A., and Smith, T. (in press). Biomass producer decision making: Direct and indirect transfers in different spheres of interaction. *Accepted for publication in Environmental Management.*

Gazzoni, D.L., Henning, A.A., Garcia, S., and Lantmann, A. (2002). Guidelines for good agricultural practices: Soybean production. In *Guidelines for good agricultural practices* (pp. 235–268). Londrina, Brazil: FAO/Embrapa.

IFA. (2012). *Fertilizer recommendations.* Retrieved 21 January 2012 from www.fertilizer.org/ifa/content/. . ./1/. . ./maize.pdf

James A. Baker III Institute for Public Policy. (2010). *Fundamentals of a sustainable U.S. biofuels policy.* Retrieved 12 October 2012 from http://www.baker institute.org/publications/EF-pub-BioFuelsWhitePaper-010510.pdf

Koplow, D. (2009). State and federal subsidies to biofuels: Magnitude and options for redirection. *International Journal of Biotechnology,* Vol 11, pp92–126.

Laborde, D. (2010). *Assessing the land use change consequences of European biofuel policies: Final report.* Washington, DC: International Food Policy Research Institute.

Maize. (2012). *Wikipedia.* Retrieved 25 November 2012 from http://en.wikipedia. org/wiki/Maize

Maize/corn: General information. (2012). *Wikidot.* Retrieved 7 October 2012 from http://cropnutrition.wikidot.com/maize-corn: general-information

Monfreda, C., N. Ramankutty, and J. A. Foley (2008), Farming the planet: 2. Geographic distribution of crop areas, yields, physiological types, and net primary production in the year 2000, *Global Biogeochem. Cycles,* 22, GB1022, doi:10.1029/2007GB002947.

OECD-FAO. (2007). *Agricultural outlook 2007–2016.* Paris, France: Organisation for Economic Co-operation and Development.

Oladosu, G., Kline, K., Uria-Martinez, R., and Eaton, L. (2011). Sources of corn for ethanol production in the United States: A decomposition analysis of the empirical data. *Biofuels, Bioproducts and Biorefining,* Vol 5, pp640–653.

Pandey, A. (Ed.). (2009). *Handbook of plant-based biofuels.* Boca Raton, FL: CRC Press.

Pradhan, A., Shrestha, D. S., McAloon, A., Yee, W., Haas, M., and Duffield, J. A. (2011). Energy life cycle assessment of soybean biodiesel revisited. *American Society of Agricultural and Biological Engineers,* Vol 54, pp1031–1039.

Sanchez, O. J., and Cardona, C. A. (2008). Trends in biotechnological production of fuel ethanol from different feedstocks. *Bioresource Technology,* Vol 99, pp5270–5295. Retrieved 11 June 2012 from http://biotecnologia.ucaldas.edu. co/publicaciones/Article-Trends%20in%20BT%20prdn%20of%20fuel%20 EtOH%20from%20different%20feedst.pdf

Schnepf, R., and Yacobucci, B. D. (2012). *Renewable fuel standard (RFS): Overview and issues.* Congressional Research Service. Retrieved 12 October 2012 from http://www.fas.org/sgp/crs/misc/R40155.pdf

Searchinger, T., Heimlich, R., Houghton, R. A., Dong, F., Elobeid, A., Fabiosa, J., Tokgoz, S., Hayes, D., and Yu, T. (2008). Use of U.S. croplands for biofuels increases greenhouse gases through emissions from land-use change. *Science,* Vol 319, pp1238–1240.

Siebert, S., Portmann, F. T., and Döll, P. (2010). Global patterns of cropland intensity. *Remote Sensing,* Vol 2, pp1625–1643.

Slingerland, S., and van Geuns, L. (2005). *Drivers for an international biofuels market.* The Hague: Clingendael Institute. Retrieved 11 July 2008 from http://www .clingendael.nl/publications/2005/20051209_ciep_misc_biofuelsmarket.pdf

Table 2: Biofuel yields for different feedstocks and countries. (2012). *Greenfacts.* Retrieved 21 October 2012 from http://www.greenfacts.org/en/biofuels/ figtableboxes/biofuel-yields-countries.htm

USDA. (2011). *Agricultural projections to 2020.* Retrieved 12 January 2012 from http://www.ers.usda.gov/Publications/OCE111/OCE111.pdf

USDA ERS. (2012). *Total U.S. jobs created by agricultural exports rises in 2011.* Retrieved 4 March 2013 from http://www.ers.usda.gov/data-products/chart-gallery/detail.aspx?chartId=35507&ref=collection

Voegele, E. (2013). Report highlights economic impact of ethanol in N.D. *Ethanol Producer.* Retrieved 22 January 2013 from http://ethanolproducer.com/articles/9477/report-highlights-economic-impact-of-ethanol-in-n-d

Waste to biofuels market analysis 2013. (2012). *Renewable Waste Intelligence.* Retrieved 15 January 2013 from http://www.renewable-waste.com/biofuels/content6.php

Wise, T. A. (2012). *The cost to developing countries of U.S. corn ethanol expansion.* GDAE Working Paper No. 12–02.

8 Biofuel production in the EU

J.W.A. Langeveld, P.M.F. Quist-Wessel
and H. Croezen

8.1 Introduction

This chapter discusses biofuel production in the EU, an economic bloc of industrial nations on the European continent. Since its inception, the EU has strongly focussed on the development of energy security and on improving coherence between the different rural regions. A strong policy stimulating farm development and agricultural output has been one of the most distinct instruments. Up to date, agriculture and rural development make up a large share of the EU budget.

This policy has been very successful, as agricultural output in the EU has shown enormous improvements. One of the downsides of this success has been, however, a decline of prices of agricultural crops following overproduction especially in the 1970s and 1980s. Several adjustments since then had to be made to the program in order to discourage increases in farming output and to limit economic support to farmers affected by declining prices. To date, remnants of this policy (e.g., a program to stimulate voluntary fallow of arable land) still are in place.

The EU has played an important role in the global development of biofuel production. Building on a desire to cut emissions of GHG, and as a way to make use of a surplus of agricultural land, it was decided in 2005 to put in place legislation of compulsory replacement of fossil transportation fuels by biofuels. It was the implementation of this legislation, and the strong response from the biofuel industry that followed, that has been one of the driving forces stimulating the biofuel development since the start of the twenty-first century.

This chapter explains what policy has been implemented and how agriculture in the EU – among other players – has responded. This is done by exploring the dynamics of land use in the EU member states (28 to date) and the way food and feed crops in their domains have been produced throughout the past decades. The chapter uses concepts that have been introduced earlier in the book. When this occurs, the reader is referred to earlier chapters.

The chapter is organised as follows. First, land cover and land use in the EU are analysed (Section 8.2). Next, it is discussed how a biofuel industry evolved in the EU (Section 8.3). Section 8.4 explains to what extent domestic crops are involved in biofuel production, while Sections 8.5 and 8.6 elaborate on the dynamics of the production of two major biofuel feedstocks in the EU (wheat and rapeseed) and their conversion to biofuels. This is followed by an evaluation of biofuel feedstock production (Section 8.7) and some conclusions (Section 8.8).

8.2 Land resources

The EU is an economic and political bloc of 28 European countries. Starting from an economic and strategic cooperation in the fields of energy (coal) and industrial (steel) production, it has developed into an economic bloc of global importance, hosting several of the largest economic powers in the world.

EU countries have a temperate climate ranging from continental in the east to the more moderate sea climates near the North Sea. In terms of land, the EU is not as large as the USA, China, or Brazil. It covers some 410 million ha of land and consists mostly of agricultural land, which covers nearly half (45%) of the area. Figure 8.1 depicts a schematic overview of different types of land use in the EU. Next to agriculture, the EU hosts forest (38%), and 'other' land (that is not used for agriculture and is not considered forest). The share of this 'other' land (18%) is quite high, although somewhat less than in the USA.

Figures on land area have been taken from the FAO (Table 8.1). Forest area covers 157 million ha and 'other' land covers 74 million ha. The remainder (187 million ha) is devoted to agriculture. This area consists of arable land (107 million ha), permanent grassland (68 million ha), and agricultural tree crops (12 million ha). While utilisation of the latter two (grassland, tree crops) is more or less permanent, farmers decide each year which crops are to be grown on the arable land. Some 80 million ha is used to grow arable crops. In 2010, 22 million ha was used to produce fodder crops; an estimated area of 6 million ha is under fallow.

These figures suggest that land use in the EU has a high share of arable crops, while area of grassland or fodder crops is relatively low. Cropping intensity, defined as the area of harvested arable crops per unit of arable land, and expressed by the MCI, is around 0.7.

Fallow and set-aside data are not easy to obtain. According to Siebert et al. (2010), who reviewed cropping intensities with a combination of crop census data and satellite images, Europe hosts some 73 million ha of fallow land. The share of fallow is similar to that of the USA. Part of the fallow land is organised under a specific scheme where farmers are compensated to leave part of their arable land under fallow. This scheme, referred to as 'set-aside', was originally introduced in the 1980s as an instrument to prevent the decline of agricultural prices due to overproduction. Set-aside area has been estimated at 7 million ha in 2007.

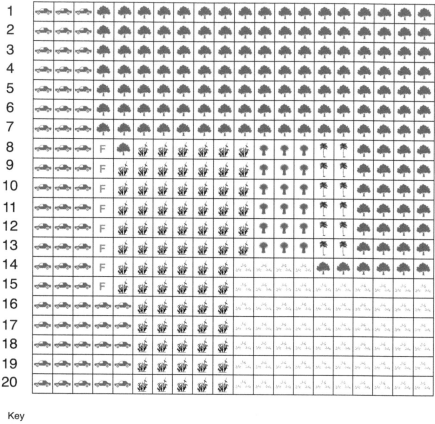

Key

🌳 forest	··· permanent grassland	🌿 arable crops			
🌴 fodder	🌴 agricultural tree crops	🚗 other	F fallow		

Figure 8.1 Schematic overview of land use in the EU

No geographical representation. Each cell represents 0.25% of land area. Position of the categories was chosen randomly.

Source: Calculated from FAOSTAT (2010–2013), faostat.fao.org

Data presented in Table 8.1 allow an analysis of land cover and land-use changes since 1980, although some data (e.g., forest and 'other' land) are not available before 1990. Agricultural area has shown a continuous decline since 1980. Arable land peaked around 2000 but has decreased since. Permanent grassland has been declining while the area of tree and arable crops is more or less stable. Consequently, loss of arable area has gone at the extent of fodder crops.

Table 8.1 Land area and land use in the EU

	Unit	1980	1990	2000	2010
Forest area	million ha	No data	136	153	157
Other land	million ha	No data	61	68	74
Agricultural area	million ha	200	195	199	187
Permanent grassland	million ha	74	71	71	68
Agricultural tree crops	million ha	14	13	13	12
Arable area	million ha	112	111	115	107
Of which:					
arable crop area	million ha	81	82	84	79
fodder crop area	million ha	31	29	24	22
fallow	million ha	No data	No data	7.6[1]	6.3
Set-aside	million ha	No data	No data	6.9[2]	
Multiple Cropping Index (MCI)	–	0.72	0.74	0.73	0.75

[1] Data for 2001.
[2] Data for 2007.

Source: Calculated from FAOSTAT (data retrieved between 2010 and 2013 from http://faostat.fao.org) and Areté and University of Bologna (2008).

Since 2000, 12 million ha of agricultural land has been converted. It must be assumed that nearly half of this has been converted into non-agricultural land ('other land', mostly related to urbanisation, etc.). During the same period, 8 million ha of arable land was lost, plus 3 million ha of grassland. Cropping intensity declined in the 1980s and early 1990s but started increasing again after 1993. Increases in cropping intensity have allowed 4 million ha of additional harvest from arable land (not requiring area expansion) since 2000. This compensates one-third of the lost agricultural area.

8.3 Markets and policy conditions

Since its inception, the EU has stimulated agricultural production through the so-called common agricultural policy (CAP). This policy has been highly successful, turning the EU from a food importer to a net exporter. On the backside, a system of food surpluses developed that had to be stored and disposed of at considerable costs. Following the increasing costs, the CAP has been gradually adjusted. One element of the adjustments was the set-aside programme, in which farmers in member states (then 15) were obliged to keep part of the arable land under fallow. Consequently, the use of surplus agricultural land to generate biomass for bioenergy seemed like a logical step that could serve both the environmental objectives (by replacing part of fossil transportation fuels) as well as farmers' interests.

The European Commission presented a Strategic European Energy Review in 2007, including a Renewable Energy Road Map to provide a long-term vision for renewable energy sources. A binding target of 20% for renewable energy's share in energy consumption in the EU was proposed by 2020, plus a 10% target for the share of renewable energy in transport petrol and diesel. Already in 2003, a directive had been accepted on the promotion of the use of biofuels or other renewable transportation. This directive set a target of a 5.75% blending of biofuels by 2010.

The Renewable Energy Directive (RED) of 2008 set an overall binding target of 20% renewable energy by 2020 plus a 10% binding minimum target for the market share of biofuels for member states in 2020. Member states were free to develop the renewable energy sector that corresponded best to their national situation and potential, provided that they collectively reach the 20% target. A plan was proposed for gradual increase of the biofuel blending share, while national action plans were requested in which member states were to define the way the final obligation was to be met. The RED also provided typical and default values for GHG emission savings by biofuel feedstock chains (often including specifications for conversion routes), as well as a formula for the calculation of emissions realised during the production and use of biofuels. Both were based on detailed research on the production and conversion of biofuel feedstocks by formal EU research institutes, work that has been published widely (JRC, 2006). Biofuels were to realise a minimum of 35% GHG reduction, a figure that was to be raised to 50% in 2017 and 60% for new installations in 2018.

The RED was followed by several publications in which its principles and calculation methods were further defined and explained. EU biofuels policy has been widely criticised for pushing up fuel and food prices, and for being an ineffective approach for reducing GHG emissions. An area of main critique has been the fact that indirect effects of enhanced biofuel production on land use (i.e., ILUC) were not accommodated. Following publication of the Searchinger et al. (2008) and Fargione et al. (2008) research papers quantifying ILUC impacts, the focus increasingly centred on defining and quantifying these impacts.

Following the discussion, in October 2012 the European Commission, after nearly four years, finally published its long expected proposal to adjust the RED and accompanying Fuel Quality Directive to accommodate ILUC caused by biofuel production (domestic as well as imported). The Commission (Holzer, 2013) decided to do the following:

- Raise the minimum GHG-saving threshold for new installations to 60%
- Impose reporting of ILUC emissions by member states and fuel distributors
- Limit ('cap') the contribution of biofuels from primary ('food') crops to 5% by 2020
- Provide further support to 'advanced' biofuels (those produced from lignocellulosic feedstocks including wastes, straw, and algae).

An overview of policy objectives is given in Table 8.2.

Table 8.2 Biofuel policy and production in the EU

Year	Instrument	Remarks
2003	Strategic European Energy Review	Promotion of biofuel use. A 5.75% blending target by 2010.
2007	Renewable Energy Road Map	Long-term vision for renewable energy, defining a 20% target for renewable energy in energy consumption by 2020, plus a 10% target for renewable energy in transport.
2008	Renewable Energy Directive (RED)	Formal implementation of the 2007 targets.
2010	Guidelines	Guidelines for the calculating of soil carbon stocks RED.
2012	Adjustment of RED and Fuel Quality Directive	Minimise impact of indirect land-use change, reformulating 2020 targets; limit contribution of food crops to biomass; provide support for biofuels from lignocellulosic biomass.

Source: European Union (2009); European Commission (2010).

Following the implementation of the biofuel policy, producers and traders developed a biofuel industry focussing on biofuel feedstock production, trade, and conversion. Domestic ethanol production increased from 1.5 billion litres in 2000 to 6.4 billion in 2010 (Table 8.3). This makes the EU a relatively small producer, lagging far behind Brazil and the USA. Europe represents the third largest ethanol market in the world, consuming 5.9 billion litres in 2010, of which some 75% was produced domestically. Notwithstanding its size, the role of ethanol markets in the USA (48.5 billion litres in 2010) and Brazil (24.8 billion litres) are much more important. A strong increase in EU ethanol production is projected, amounting to 18 billion litres in 2019 (OECD-FAO, 2010).

Biodiesel production showed an impressive growth from 800 million litres in 2000 to 10 billion litres in 2010. Consumption showed a similar increase but is still exceeding domestic production by some 35%. The Food and Agricultural Policy Research Institute, Iowa State University (FAPRI-ISU, 2011) is projecting a further increase for biofuel production and consumption up to 2020. According to the OECD-FAO (2010), biodiesel production could reach 24 billion litres in 2019.

Ethanol production costs declined by one-third since 2000 but still were very high in 2010 at US$0.63, and a considerable levy (40% of production costs) is imposed on ethanol imports. For the near future, further declines in production costs are expected (Steenblik, 2007). In addition, some member states offer considerable subsidies or tax exemptions to ethanol producers. The data given in Table 8.3 suggest that locally, high shares of the production

Table 8.3 Biofuel market development and economics in the EU

	Unit[1]	Bioethanol		Biodiesel	
		2000	*2010*	*2000*	*2010*
Biofuel production	billion litre/year	1.5	6.4	0.8	10.3
Biofuel consumption	billion litre/year	1.7[2]	5.9	3.5[3]	14.0
Support investments	billion $	No data	1.2[4]	No data	No data
Subsidies, tax exemptions	US$/litre	1.25[5]	0.02–0.85		0.04–5.35
Subsidies, tax exemptions (average)			0.20–0.28		0.43–0.52
Import levies	US$/litre		0.24		
Production cost	US$/litre	0.90[4]	0.63		

[1]Billion = thousand million.
[2]Data for 2006.
[3]Data for 2005.
[4]Data for 2007.
[5]Data for France.
[6]Data for 2011.

Source: OECD (2008); OECD-FAO (2012); FAPRI-ISU (2011); Popp (2009); Steenblik (2007); GSI-IISD (2013).

costs may be covered by subsidies. Member states and the EU further provide support to cover high investment costs. Total biofuel support in the EU (investment, other subsidies, tax exemptions for bioethanol and biodiesel) has been quantified at US$7.4 to 9.3 billion in 2011 (GSI-IISD, 2013), which is high when compared to support in the USA.

8.4 Feedstock requirements

Wheat and rape are the dominant feedstocks used in biofuel production in the EU. Ethanol production in Europe is mostly grain based. Wheat is the most important ethanol feedstock, and ethanol production in Europe is dominated by EU member states. Production capacity in the EU amounted to 8.5 billion litres in 2012, half of which was in use ('Annual EU Report', 2013). France and Germany dominate (annually producing 1.8 and 1.2 billion litres in 2009, respectively), followed by Spain (560 million litres), the Netherlands (515 million litres), Belgium, and the UK.

Statistics published by ePURE, the European renewable ethanol association, show that Germany is home to 10 ethanol plants representing a total capacity of more than 1.2 billion litres. According to ePURE data, Germany produced 760 million litres of ethanol in 2010. In 2012, 40% of the ethanol was manufactured from sugar beet feedstock. Feed grain was used to produce the remainder ('Sugar Beet Feedstock Use', 2013).

Rapeseed is the most common feedstock for EU biodiesel production, representing nearly 90% of all feedstocks in 2010 (FAPRI-ISU, 2011). This crop was already known in central European countries such as Germany, the Ukraine, and Poland, and recently has become more common in France and the UK. Following the demand for biodiesel, rapeseed cultivation in these countries increased quickly along with further development of conversion capacity, especially in Germany and Spain.

Wheat to ethanol

Wheat is the second most important cereal crop in the world (after corn), and its importance is increasing. Annual production is expected to rise from 650 million tonnes to 900 million tonnes in 2050, showing a stronger growth than rice but not as strong as corn, which is projected to double in the period from 2000 to 2050.

In Europe wheat is the major food crop. Its cultivation amounts to some 60 million ha, a one-third decline since 1980. Notwithstanding this area decline, wheat *production* showed a strong yield increase (+64% since 1980, representing an average annual increase of more than 2% per year) to 225 million tonnes (data retrieved between 2010 and 2013 from http://faostat.fao.org). European wheat area represents about one-quarter of total cultivated area in the world, and yields in this continent are among the highest globally. The Russian Federation and the Ukraine are major players with 22 and 6 million ha, respectively. Main producers in the EU include France (5 million ha) and Germany (3 million ha), Poland, Romania, the UK, and Spain.

Wheat area in the EU increased by 4 million ha in 1980 to 27 million ha in 2000, and it has remained constant since then (Table 8.4). It is projected to increase to 28 million ha in 2020. Wheat yields have been increasing since the 1950s. In the past two decades of the twentieth century, annual yield

Table 8.4 Wheat production in the EU

	Unit	1980	1990	2000	2010	2020
Harvested area	million ha	23	24	27	27	28
Production	million tonne	88	117	132	139	155
Yield (three-year average)	tonne/ha	3.6	4.8	4.9	5.4	5.5
Average annual increase[1]	kg/ha/year		118	13	59	10

[1]Figures for 1980–1990 are presented in the column for 1990, figures for 1990–2000 are presented in the column for 2000, and so on.

Source: Calculated from FAOSTAT (2010-2013); http://www.faostat.fao.org; FAPRI-ISU (2011).

improvement amounted to 3.4%. This is one of the highest figures world-wide. Since 2000, annual yield increase in Europe has been modest at 1.3% per year. Average annual yield increase has been variable: almost 120 kg in the 1980s, which declined to 13 kg in the 1990s and has since then almost quadrupled to some 60 kg per ha per year.

The use of cereals for biofuel production in Europe has only recently developed. According to OECD-FAO (2007), production started to take off in 2003. Total cereal use remained below 5 million tonnes until 2006. Wheat started to dominate after 2006, reaching a consumption level of around 7 million tonnes in 2010. This represents 5% of domestic wheat production (Table 8.5). Biofuel wheat area in the EU is 1.8 million ha.

Wheat use in ethanol production has been projected to rise to 17 million tonnes, but this may be on the high side as this projection was done before the latest policy change in 2012. This amount of wheat would require 3.9 million ha (FAPRI-ISU, 2011).

So far, there seems to be little difference between wheat production for food (feed) and cultivation oriented primarily towards biofuel feedstocks. For the UK, biofuel yields were reported at 7.5 tonnes per ha, which seems to be in line with wheat yields for other purposes (HGCA, 2010). Expectations for yield increases in the UK are quite optimistic, with an average annual increase of some 150 kg per ha per year up to a yield level of 9 tonnes per ha.

Rape to biodiesel

Oilseed rape (canola), an annual C3 crop grown in temperate areas, is of increasing importance. The seeds contain 40% oil that is used for salad oil, as a cooking medium, and for the production of biofuel. The residues referred to as oilseed cake are rich in protein and are used as animal feed. In many parts of South Asia (including India), rapeseed mustard is an important winter crop that is grown either alone or as a secondary intercrop in wheat fields. New varieties of winter rape, including hybrids, can attain high

Table 8.5 Wheat use for ethanol production in the EU

	Unit	2000	2010	2020
Use in biofuels	million tonne	Negligible	7.0	17.0[1]
Share of all wheat	%	Negligible	5%	7%
Biofuel feedstock harvested area	million ha	Negligible	1.8	3.9

[1]Figure for 2016.

Source: Calculated from OECD-FAO (2007); FAPRI-ISU (2011); FAOSTAT (2010-2013); http://www.faostat.fao.org; FAO (2008); Burrell (2010); BioGrace (2012).

Table 8.6 Rapeseed production in the EU

	Unit	1980	1990	2000	2010
Harvested area	million ha	1.5	3.0	4.1	6.9
Production	million tonne	3.6	8.6	11.3	20.4
Yield (three-year average)	tonne/ha	2.4	2.9	2.8	3.1
Average annual increase[1]	kg/ha/year	–	63	–11	31

[1]Figures for 1980–1990 are presented in the column for 1990, figures for 1990–2000 are presented in the column for 2000, and so on.

Source: Calculated from FAOSTAT (2010-2013); http://www.faostat.fao.org.

Table 8.7 Rapeseed use for biodiesel production in the EU

	Unit	2000	2005	2010	2020
Use in biofuels	million ton	6	7	20	25
Share of all rapeseed	%	55%	43%	100%	100%
Biofuel feedstock harvested area	million ha	2.1	2.1	6.9	7.1

Source: Calculated from OECD-FAO (2007); FAPRI-ISU (2011); FAOSTAT (2010-2013); http://www.faostat.fao.org.

seed yields of 4–5 tonnes per ha as compared to current yield levels, which are 3–3.5 tonnes per ha in Europe (FAO, 2012).

Rape area in the EU quadrupled from 1.5 million ha in 1980 to almost 7 million ha in 2010 (Table 8.6). During this period, production increased more than fivefold to 20 million tonnes. EU production is mainly concentrated in France and Germany, each hosting 1.5 million ha in 2010. Rape yields rose until 1990, then declined to reach the 1990 level again in 1997. These levels were surpassed in 2004, and they have shown a continuous increase. Since 1980 yields have been up by almost 50%. Average annual increase in the 1980s was more than 60 kg per ha per year. Following yield declines, the increase since 2000 has been 31 kg per ha per year.

The use of rapeseed in the biodiesel industry tripled in five years to 20 million tonnes in 2010 (Table 8.7). It is projected to grow further, but following recent policy changes, the actual increase may be less than the 25 million tonnes predicted for 2020. In 2010 all rapeseed was used to produce biofuels. This makes the biofuel area for rapeseed some 7 million ha, up by 5.8 million ha since 2005.

8.5 Crop cultivation

Crop cultivation in the EU is mainly based on three- and four-year crop rotations to maintain soil fertility and limit the occurrence of pests and diseases. Monocultures are found locally in the north while monocultures and short

Year	Italy (1)	Italy (2)	Italy (3)	Spain (1)	Spain (2)	Hungary	France (1)	France (2)	Austria / France
1	Soybean	Fava bean	Alfalfa	Fava bean	Fallow	Peas	Pea	Pea	Pea
2	Wheat	Wheat	Alfalfa	Wheat	Wheat	Wheat	Wheat	Wheat	Pea
3			Alfalfa				Maize or barley	Maize or barley	Wheat
4			Wheat or maize						Oilseed rape
5									Barley

Figure 8.2 Typical short wheat rotations in Europe
Source: Mudgal et al. (2010).

rotations are reported in the Mediterranean regions (Mudgal et al., 2010). Common wheat rotations are presented in Figure 8.2. Short rotations generally include wheat every second or third year. Other crops are generally pulses, other cereals, or fodder crops.

Wheat

Wheat is an annual cereal C3 crop. It can be sown in spring ('summer wheat') or in autumn ('winter wheat'). The latter is most common in the EU. A typical crop calendar for winter wheat is presented in Figure 8.3.

Wheat in France is planted from the beginning of October to the end of November. The harvest season begins around July 1 and is usually concluded by the end of August. The planting season for wheat in Germany roughly covers the month of October. Harvest begins the next August and runs through the end of the month (Spectrum, 2013). Winter wheat in the UK is planted around October 1, with planting running through the beginning of December. Harvest normally begins around July 1 and concludes at the end of August.

Activity	Jan	Feb	Mar	Apr	May	Jun	Jul	Aug	Sep	Oct	Nov	Dec
Sowing								X	X			
Growing period	X	X	X	X	X	X	X			X	X	X
Harvesting						X	X	X				
Post-harvest								X	X			

Figure 8.3 Cropping calendar for winter wheat in the EU.

Wheat prefers fertile soils but can be grown in many soil types, except very light sandy soils or peat soils, provided the water requirement and nutrient demand can be met (Burrell, 2010). Good yields, however, require fertile soils with good structure and porous subsoil for deep roots. The optimal soil is slightly acid to neutral. The water supply should not be restrictive, and rains should be well distributed (FAO, 2006). Ploughing is common before sowing, especially when animal manure is applied. Straw is removed after harvest and sold. Sometimes, a catch crop is sown after harvest to make use of available mineral nutrients plus additional nutrients that are released after the wheat is harvested. In this way, catch crops incorporate nutrients into organic matter, preventing them from immediate leaching to ground or surface water.

Data on input use for wheat cultivation are presented in Table 8.8. Artificial nitrogen fertiliser applications range from 60 to 200 kg of nitrogen (N) per ha. Applications should be related to soil fertility and mineralisation of previous crop residues. Typically, around 130 kg per ha of nitrogen is applied. Phosphorus (P) applications, ranging between 30 and 120 kg of phosphate per ha, would typically amount to 60 kg P_2O_5. The range of potassium (K) applications (60–240) is larger than phosphorus. A typical application in Germany would be 120 kg of K_2O per ha (Heyland et al., 2013). Generic data on the application of animal manure is scarce. Manure applications have been reported in the Netherlands and Germany and are further expected in regions with high animal density.

Table 8.8 Wheat input use in the EU

Parameter (unit)	Per ha	Source
Nutrient application (kg/ha)		
N	60	Germany (Heyland et al., 2013)
	80–185	Western Europe (Isherwood, 2012)
	109	BioGrace (2012)
	80–200	Typical values
P_2O_5	60	Heyland et al. (2013)
	22	BioGrace (2012)
K_2O	120	Germany (Heyland et al., 2013)
	16	BioGrace (2012)
Application of agro-chemicals	3.5	Webb et al. (2010)
(kg active ingredient/ha)	1.2	BioGrace (2012)
Diesel (litre/ha)	90–170	Typical values
	112	Webb et al. (2010)
	100	BioGrace (2012)
Required rainfall (mm)	250–1,750	Curtis (2002)

Cool and moist conditions in the northern parts of the EU are condu-cive to diseases, which require multiple fungicide applications. The reported applications of agro-chemicals in the UK (3.5 kg of active ingredients per ha) (Webb et al., 2010; Te Buck et al., 2010) are quite high. The reported diesel use is 112 litres per ha in the UK (Webb et al., 2010). Higher figures have been reported for the Netherlands (176 litre per ha) (Te Buck et al., 2010) while lower data (less than 100 litre per ha) were given for France and Germany.

Rapeseed

Rapeseed is a mustard-like annual oil crop related to the cabbage family. It is typically grown in cooler summer climates, such as in the northern USA, Canada, and Germany, and produces fields of colourful yellow flowers containing seeds with a high oil content and unique fatty acid composi-tion (Drapcho et al., 2008). It can be sown in spring ('summer rape') or in autumn ('winter rape'). This book focuses mostly on the winter type. A typical crop calendar is presented in Figure 8.4. Sowing starts in August and may continue in September. Harvesting is done in the following summer (June–August). It may be followed by sowing of a catch crop.

Rape is a demanding crop that requires a deep fertile topsoil without com-pacted layers (to facilitate root growth) and a porous crumb structure of the uppermost soil layer (for rapid germination of the seed). It prefers a deep, sandy loam rich in humus and nutrients and with an optimal lime content. Alternatively, soils could be humic loams, loam soils, or clayey loam soils (in decreasing order of suitability). Humic loams and loamy soils require sufficient precipitation in April. Summer rape has a weaker root system than winter rape, which makes it more sensitive to water deficit. Marshy or waterlogged soils are unsuitable (FAO, 2006; El Bassam, 2010).

Ploughing is common before sowing, especially when animal manure is applied. Straw is usually not removed after harvest.

Artificial nitrogen fertiliser applications range from 150 to 250 kg of nitro-gen (Table 8.9). Typically, around 200 kg of nitrogen would be applied per ha. Phosphorus applications, ranging between 80 and 100 kg of phosphate

Activity	Jan	Feb	Mar	Apr	May	Jun	Jul	Aug	Sep	Oct	Nov	Dec
Sowing								X	X			
Growing period	X	X	X	X	X	X	X			X	X	X
Harvesting						X	X	X				
Post-harvest								X	X			

Figure 8.4 Cropping calendar for rapeseed in the EU

Table 8.9 Rapeseed input use in the EU

Parameter (unit)	Per ha	Source
Nutrient application (kg)		
N	150–250	Typical value
	200	FAO (2006)
P_2O_5	80–100	Typical value
	90	FAO (2006)
K_2O	80–220	Typical value
	200	Germany (FAO, 2006)
Application of agro-chemicals (kg of active ingredient/ha)	1.2	BioGrace (2012)
	3.0	Webb et al. (2010); Te Buck et al. (2010)
Diesel (litre/ha)	84–172	BioGrace (2012)
	83	Webb et al. (2010); Te Buck et al. (2010)
Required rainfall (mm)	400–500	

Source: FAO (2008).

per ha, would typically amount to 90 kg P_2O_5. The range of potassium applications (180–220) is larger than phosphorus. A typical application in Germany would be 200 kg of K_2O per ha (Heyland et al., 2013). Animal manure can be applied when available. Application of agro-chemicals in the UK (Webb et al., 2010) amounts to 3 kg of active ingredients per ha, which is rather high. The average diesel use reported for EU bioenergy cropping systems varies from 84 to 172 litres per ha (e.g. Webb et al., 2010; Te Buck et al., 2010), with the highest figures reported for the Netherlands.

8.6 Conversion to biofuel

Wheat to ethanol

The basic conversion route for wheat ethanol is presented in Figure 8.5. It contains two major steps: (1) production of sugar from starch and (2) fermentation of the sugar to ethanol. Sugar is made through milling of the grains, dilution, and heating of the solution to dissolve the starch, which then is converted into sugars by hydrolysis (De Wit et al., 2010). Hydrolysis may include acid and enzymatic methods. Acid hydrolysis results in the production of unnatural compounds that have adverse effects on yeast fermentation, whereas enzymatic hydrolysis requires cooking biomass at high temperatures and adding large amounts of enzymes. Because this is costly, alternative approaches are being developed (Pandey, 2009), and more efficient approaches may be expected in the near future. Distillation of the fermented sugar solution is an energy-intensive step that requires

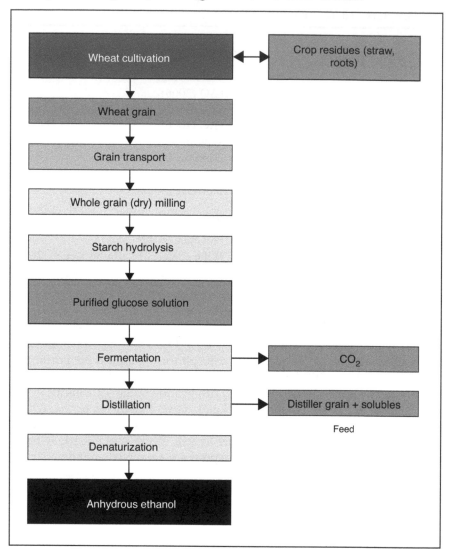

Figure 8.5 Conversion from wheat to ethanol. Figure design by H. Croezen.

considerable amounts of fuels or electricity. The climate effect of the energy source will therefore have a large impact on the GHG-reducing potential of the ethanol (see Langeveld and van de Ven, 2010).

Ethanol contains just more than half of the feedstock energy, with the remainder being recovered in the co-product DDGS (Table 8.10). Wheat to ethanol conversion is 373 litres per tonne of wheat, which yields some 1,900 litres of ethanol per ha per year. Conversion efficiency has been projected to improve by 0.5% per year (Burrell, 2010), which is equivalent to 1.9 litres

Table 8.10 Efficiency of wheat-to-ethanol conversion in the EU

	Unit	Biograce	Other Sources	
			2010	2020
Conversion efficiency	litre/tonne		373	392
Conversion efficiency	GJ/GJ	0.52		
Biofuel yield	litre/ha		1,900	
Biofuel yield	GJ/ha	76.6		
Co-product yield	tonne/ha	2.8	2.7	

Source: Calculated from BioGrace (2012); Elsayed et al. (2003); Burrell (2010).

per tonne of wheat per year. We calculated an efficiency of 392 litres per tonne in 2020 (annual improvement of 0.5% to the 2010 figure). The theoretical ethanol yield maximum is 550 litres per tonne (Drapcho et al., 2008).

Rape to biodiesel

The conversion route for biodiesel production from rapeseed is presented in Figure 8.6. The process consists of the following three steps. (1) Seeds are pressed into oil plus a press cake, conserving about two-thirds of the energy in the oil. (2) The oil is refined and submitted to esterification to produce methyl esters. In this process, it reacts with light alcohols (typically methanol or ethanol) at atmospheric pressure and temperatures of about 70°C. Cheap bases serve as catalysts. (3) Finally, esterification is applied to lower the viscosity of the oil. It is done with ethanol as it has more favourable environmental features. The technology is well known and has been optimised for biodiesel production since the early 1990s (Pandey, 2009); however, prospects for efficiency improvement are limited.

Conversion efficiency of rape biodiesel is 400–440 litres of biodiesel per tonne of seeds (Table 8.11), which is quite high. Rapeseed gives a higher biodiesel yield than soybean because it contains more oil: 37 kg oil per kg seed compared to 14 kg oil per kg for soybean (Drapcho et al., 2008). Nearly 60% of the energy is recovered in biodiesel. Co-products are press cake and glycerine. Following BioGrace, average biodiesel yield for the EU is nearly 1,300 litres per ha. This is confirmed by other sources. We used data provided by Stephenson et al. (2008), who studied biofuel chain systems in the UK, to calculate conversion efficiency at 441 litres of biodiesel per tonne of seed and a yield of nearly 1,400 litres of biodiesel per ha. Other sources (e.g., Elsayed et al., 2003), however, are not so optimistic.

According to Burrell (2010), conversion efficiency is not expected to show strong improvement during the upcoming few years. We calculated an efficiency of 439 litres per tonne in 2020 (an improvement of 5% over the average 2010 figure).

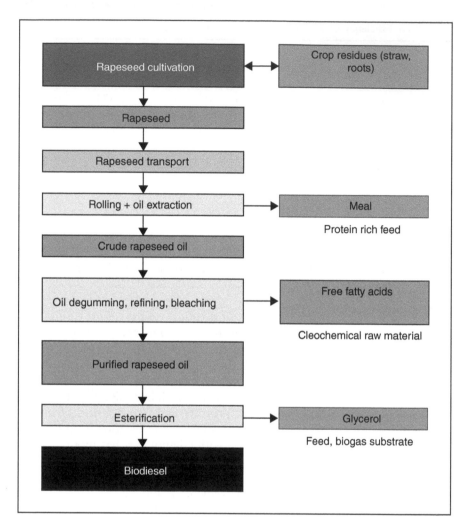

Figure 8.6 Conversion from rapeseed to biodiesel. Figure design by H. Croezen.

Table 8.11 Efficiency of rapeseed-to-biodiesel conversion in the EU

	Unit	*BioGrace*	*Other Sources*	
			2010	*2020*
Conversion efficiency	litre/tonne	416	396–441	439
Conversion efficiency	GJ/GJ	0.58	0.55–0.61	
Biofuel yield	litre/ha	1,300	1,100–1,400	1,500
Biofuel yield	GJ/ha	49	36–46	50
Co-product yield	tonne/ha	1.7		

Source: Calculated from BioGrace (2012); Stephenson et al. (2008); Elsayed et al. (2003); Burrell (2010).

8.7 Evaluation

Biofuel production

Since the implementation of the biofuel policy, the EU has experienced a rapid development of the biofuel industry; biodiesel production in particular earned an important position. However, production still lags behind consumption. Ethanol is imported mainly from Brazil, while biodiesel imports vary. High production costs remain a major obstacle for the domestic industry, which is mainly explained by high feedstock production costs while investment subsidies are lower than levels reported for Brazil and the USA. A public debate on the sustainability performance of biofuels and long-awaited policy adjustments (addressing the issue of ILUC) have further added to an atmosphere of uncertainty.

The EU has a strong tradition in intensive agricultural production and consequent high input use and yields. Wheat is one of the most important arable crops. It has high yields, and domestic production has been sufficient to support the emerging ethanol industry. In 2010, 7 million tonnes of wheat was used for ethanol production, covering less than 2 million ha. This coincides with a small increase in wheat cultivation area and a revival of yield improvement, be it that the latter still lags behind improvements realised in the past.

Rapeseed is grown in the same crop rotations as wheat, making use of the same soils and under similar conditions. Given its high nutrient requirements, fertiliser applications often are twice as high, as is the case for wheat. Moreover, rapeseed straw is higher in nutrients and is easily degraded, thus releasing more nutrients that can be lost after harvest. Growing a catch crop can help limit emissions, but it is not well known how often this is done in practice. Because rapeseed area is almost entirely devoted to biofuel production (7 million ha), this requires further attention.

Changes in land use

Since 1990 cropping intensity increased by an average of 1.7% per year. This may not seem like much, but it applies to a large area of arable land (109 million ha in 2009). Since 2000, nearly 4 million ha of additional harvested area has been generated.

Changes in land for food and feed production availability in the EU since 2000 are presented in Table 8.12. Biofuels used a harvested area of 6.6 million ha. Further changes in land availability are related to loss of agricultural land (−11.5 million ha). Total losses amount to 14.9 million ha, of which one-quarter is associated with biofuel production. Losses are partly compensated by an improvement of the MCI, generating a harvested area of 3.6 million ha. The balance is a loss of 11.2 million ha.

Table 8.12 Changes in EU land availability for food and feed production 2000–2010

	Unit	Impact on land availability	Share of land loss
Increase in biofuel production	million ha	−6.6	
Allocation to co-products	million ha	3.2	
Net biofuel land use	million ha	−3.4	23%
Loss of agricultural area	million ha	−11.5	77%
Loss of arable land total	million ha	−14.9	100%
Change in MCI	million ha	3.6	
Net land balance	million ha	−11.2	

Source: Calculated from Tables 8.1, 8.5, 8.7, and FAOSTAT (2010-2013); http://www.faostat .fao.org.

Production efficiency

Historically, EU wheat cultivation received considerable nitrogenous fertiliser inputs, which can be explained by the demand for high protein (gluten) crops favourable for food application (bread). Following a biofuel wheat industry development, we may expect lower fertilisation application levels as it leads to high gluten concentrations in the DDGS, which are less desirable with respect to its application as animal feed (Drapcho et al., 2008). This will lead to an improvement of existing nutrient balances. Currently, wheat production leads to relatively small surpluses of nitrogen and phosphate while potassium surplus is quite high.

Biofuel yields are more than 1,700 litres of ethanol from wheat, and nearly 1,300 litres of biodiesel from rape (Table 8.13). Co-product yields are modest at 2.7 and 1.7 tonnes per ha, respectively. Nutrient productivity is very modest for nitrogen and is modest to fair for phosphorus. The risk of nutrient leaching is rather high. The increase of wheat and rape cultivation may, further, have implications for nutrient use and emissions. Fertilisation levels of both crops are not much different from those of other arable crops that may be replaced. FAO statistics suggest the replacement of other cereals (mainly barley and corn). Most area will, however, come from fodder crops on arable land. In 2010 fodder crop area was 2.3 million ha smaller than was the case in 2000.

With respect to the impact on water use and emissions to groundwater, wheat and rapeseed may require more water and may cause more emissions to groundwater than fodder crops. Wheat has a rather large water footprint. Water consumption, according to Gerbens-Leenes et al. (2009), amounts to 210 m^3 per GJ (at 1,900 litres of ethanol per ha). However, most of this refers to irrigation that is applied in the southern EU in a minority of the cases. As to the contribution to soil organic matter, rapeseed appears to perform in line with the average of alternative crops (Kamp et al., 2012).

Table 8.13 Performance and impacts of wheat and rapeseed biofuel chains in the EU

	Unit	Wheat	Rapeseed
Biofuel yield	litre/ha	1,700	1,300
Biofuel yield	GJ/ha	37	43
Co-product yield	tonne/ha	2.7	1.7
Digestibility of co-product	–	Average	High
Nutrient productivity (biofuel)	GJ/kg N	0.5	0.4
	GJ/kg P_2O_5	1.0	0.8
Water requirement (biofuel)	m^3/GJ	210	100
Risk of nutrient leaching	–	High	High
Impact on soil organic matter	–	Negative	Negative

GHG emissions

Formal GHG emissions related to biofuel production in the EU are presented in Table 8.14. Emissions caused by the production of fuels (crop cultivation, transport to the plant, conversion) are 33–69 and 50–52 g of CO_2-eq/MJ, respectively. This results in emission reduction by replacing biofuels at 23–63% for wheat ethanol and 40–44% for rape biodiesel. Reported ILUC emissions are highly variable, in some cases reducing ethanol reductions to nearly zero while total rapeseed biodiesel emissions may exceed those of its fossil counterpart.

Economic and social impacts

No major economic or social impacts have been reported, although enhanced demand for biofuels (including biogas feedstock production) has led to local increases of land prices (e.g., in Germany). The continent, however, is very

Table 8.14 Recent estimates of GHG emissions EU biofuel chains

	Unit	Wheat	Rapeseed
Biofuel production	g CO_2-eq/MJ	33–69	50–52
Reduction[1]	%	23–63%	40–44%
ILUC	g CO_2-eq/MJ	3–14	5–55
Total	g CO_2-eq/MJ	60–81	58–107

Note: Biofuels for domestic use.
[1]ILUC-related emissions not included. Calculated with a fossil reference of 90 g CO_2-eq/MJ.

Source: Calculated from BioGrace (2012); European Union (2009); Laborde (2010); Darlington et al. (2013); CARB (2012); Searchinger et al. (2008).

diverse, and existing extensive production practices in the east have not been affected much. Costs involved in supporting biofuel feedstock production and conversion are considerable but remain within frameworks set out by existing support programs for agriculture or (fossil) energy production.

Changes in biomass availability

Since the introduction of the biofuels policy, agricultural production in the EU has changed. Annual biomass use for biofuels in 2010 amounts to 7 million tonnes of wheat, 10 million tonnes of sugar beets, and 20 million tonnes of rapeseed. Land use and biomass production are dynamic processes that require day-to-day decision making by farmers and other actors involved. They respond to changes in the weather as well as to changes in economic and market conditions. The dynamic character of these processes is difficult to assess.

Production of arable crops per ha of arable land (i.e., BP) has increased. Output improved nearly 20% from 3.9 tonnes per ha in 1994 to 4.6 tonnes in 2009. Since the late 1990s, average output per ha showed an average annual increase of 2.5%. Consequently, each year an average of 2.7 million tonnes of biomass is produced from the same amount of arable land. Since 2000 an additional biomass output of 122 million tonnes has been generated.

Table 8.15 gives an overview of other changes in crop production since 2000. Biofuel production claimed 31 million tonnes of biomass. One-third of this is allocated to the generation of co-products. The net loss is 20 million tonnes. In total, the EU biomass output for food and feed markets declined with 51 million tonnes of biomass since 2000. Some 40% of this is related to enhanced biofuel production.

Table 8.15 Changes in biomass availability in the EU 2000–2010

	Unit	Impact on Biomass Availability	Share of Biomass
Increase total biomass production	million ton	−31	
Biomass used in biofuel production	million ton	−31	100%
Generation of co-products	million ton	11	35%
Net impact of biofuels	million ton	−20	65%
Net changes in biomass availability	million ton	−51	
Increase in Biomass Productivity (BP)	million ton	112	

Notes: Reduction of availability is depicted by negative figures. Positive figures indicate a higher availability.

Source: Calculated from Table 8.5, Table 8.7 and FAOSTAT (2010–2013), http://faostat.fao.org.

8.8 Conclusion

Biofuel production in the EU has not shown similar growth rates to those observed in Brazil or recently in the USA. Wheat, the dominant ethanol feedstock, is the major cereal in the crop rotation, and its application in biofuels so far has been modest. Rapeseed production for biodiesel has shown a strong increase, but this has remained within existing crop rotation opportunities. Impacts on land use and biomass availability have been modest to low. Improvement of MCI has been sufficient to generate the harvest of additional required area. Intensive cultivation practices of wheat and rapeseed, however, put pressure on land and water quality, and the adjustment of fertiliser applications should be considered.

References

Annual EU report projects increased ethanol production. (2013). *Ethanol Producer Magazine*. Retrieved 15 September 2013 from http://www.ethanolproducer.com/articles/10179/annual-eu-report-projects-increased-ethanol-production

Areté and University of Bologna. (2008). *Evaluation of the set aside measure 2000 to 2006*. Retrieved 12 November 2012 from http://ec.europa.eu/agriculture/eval/reports/setaside/fulltext_en.pdf

BioGrace. (2012). Retrieved 12 May 2012 from http://www.biograce.net/content/ghgcalculationtools/calculationtool

Burrell, A. (Ed.). (2010). *Impact of EU biofuel target on agricultural markets and land use*. Retrieved 12 January 2012 from http://publications.jrc.ec.europa.eu/repository/bitstream/111111111/15287/1/jrc58484.pdf

CARB. (2012). *Carbon intensity lookup table for gasoline and fuels that substitute for gasoline*. Retrieved 21 September 2013 from http://www.arb.ca.gov/fuels/lcfs/lu_tables_11282012.pdf

Curtis, B.C., Rajaram, S., and Gómez Macpherson, H. (eds.). Bread wheat. Improvement and production. (2012). Rome, Italy. Food and Agricultural Organization of the United Nations.

Darlington, Th., Kahlbaum, D., O'Connor, D., and Mueller, S. (2013). *Land use change greenhouse gas emissions of European biofuel policies utilizing the Global Trade Analysis Project (GTAP) model*. Macatawa, USA: Air Improvement Resource.

De Wit, M., Junginger, M., Lensink, S., Londo, M., and Faaij, A. (2010). Competition between biofuels: Modelling technological learning and cost reductions over time. *Biomass and Bioenergy*, Vol 34, pp203–217.

Drapcho, C.M., Nhuan, N.P., and Walker, T.H. (2008). *Biofuels engineering process technology*. New York: McGraw-Hill.

El Bassam, N. (2010). *Handbook of bioenergy crops: A complete reference to species, development and applications*. London: Earthscan.

Elsayed, M.A., Matthews, R., and Mortimer, N.D. (2003). *Carbon and energy balances for a range of biofuels options*. Sheffield, UK: Sheffield Hallam University.

European Commission. (2010). Common decision of 10 June 2010 on guidelines for the calculation of land carbon stocks for the purpose of Annex V to Directive 2009/28/EC. *Official Journal of the European Union*, Vol L 151, pp19–42.

European Union. (2009). *Directive 2009/28/EC of the European Parliament and of the council on the promotion of the use of energy from renewable sources amending and subsequently repealing Directives 2001/77/EC and 2003/30/EC.* Brussels, Belgium: European Commission.

EUROSTAT. (2012). *Crop production statistics at regional level.* Retrieved 12 February 2012 from http://epp.eurostat.ec.europa.eu/statistics_explained/index.php/Crop_production_statistics_at_regional_level

FAO. (2006). *Nutrient management guidelines for some major field crops.* Rome, Italy: Food and Agricultural Organization of the United Nations.

FAO. (2008). *The state of food and agriculture: Biofuels – prospects, risks and opportunities.* Rome, Italy: Food and Agricultural Organization of the United Nations.

FAO. (2012). *Nutrient management guidelines for some major field crops.* Retrieved 1 October 2012 from ftp://ftp.fao.org/docrep/fao/009/a0443e/a0443e04.pdf

FAPRI-ISU. (2011). *World agriculture outlook.* Retrieved 18 May 2012 from http://www.fapri.iastate.edu/outlook/2011/

Fargione, J., Hill, J., Tilman, D., Polasky, S., and Hawthorne, P. (2008). Land clearing and the biofuel carbon debt. *Science,* Vol 319, pp1235–1238. DOI: 10.1126/science.1152747

Gerbens-Leenes, W., Hoekstra, A.Y., and Van der Meer, T.H. (2009). The water footprint of bioenergy. *Proceedings of the National Academy of Science,* Vol 106, pp10219–10223.

GSI-IISD. (2013). Addendum to Biofuels—At What Cost? A review of costs and benefits of EU biofuel policies. Winnipeg, MA Canada: International Institute for Sustainable Development.

Heyland, K.-U., Werner, A., Zongheng, L., Zengshou, Y., Prasad, R., Halvorson, A.D., and Jürgens G. (2013). *Wheat.* Retrieved 15 September 2013 from http://afghanag.ucdavis.edu/b_field-crops/wheat-1/FS_Wheat_Fert_WFUM_IFA.pdf

HGCA. (2010). *Growing wheat for alcohol/bioethanol production.* Retrieved 26 September 2012 from http://publications.hgca.com/publications/documents/Biofuels.pdf

Holzer, M. (2013). *The common agricultural policy and feedstock mobilisation.* Presentation given at the Fifth EBTP Stakeholder Plenary Meeting, "Advanced Biofuels Deployment – Investing in Europe's Future," the Diamant Conference Centre, Brussels, Belgium, 6–7 February.

Isherwood, K.F. (2012). *Fertilizer use in western Europe: types and amounts.* EOLSS. Retrieved 30 September 2012 from http://www.eolss.net/sample-chapters/c10/E5–24–08–03B.pdf

JRC. (2006). *Well-to-wheels analysis of future automotive fuels and powertrains in the European context.* Well-to-tank report, Concawe: EUCAR, JRC. Retrieved 11 October 2006 from http://ies.jrc.ec.europa.eu/WTW

Kamp, J.A.L.M., de Visser, C.L.M., Hanse, B., Huijbregts, A.W.M., Meuffels, G.J.H.M., van der Voort, M.P.J., and Stilma, E. (2012). *Energieboerderij eindrapportage.* Lelystad, the Netherlands: Praktijkonderzoek Plant en Omgeving.

Laborde, D. (2011). *Assessing the land use change consequences of European biofuel policies: Final report.* Washington, DC: International Food Policy Research Institute.

Langeveld, J.W.A., and van de Ven, G.W.J. (2010). Principles of plant production. In H. Langeveld, J. Sanders, and M. Meeusen (Eds.), *The biobased economy: Biofuels, materials and chemicals in the post-oil era.* London: Earthscan.

Mudgal, S., Lavelle, P., Cachia, F., Somogyi, D., Majewski, E., Fontaine, L., Bechini, L., and Debaeke, Ph. (2010). *Environmental impacts of different crop rotations in the European Union*. Paris, France: Bio Intelligence Service.

OECD. (2008). *Economic assessment of biofuel support policies*. Retrieved 19 June 2012 from http://www.oecd.org/dataoecd/54/10/40990370.pdf

OECD-FAO. (2007). *Agricultural outlook 2007–2016*. Paris, France: Organisation for Economic Co-operation and Development.

OECD-FAO. (2010). *Agricultural outlook 2010–2019*. Paris, France: Organisation for Economic Co-operation and Development.

OECD-FAO. (2012). Agricultural outlook 2012–2021. Paris, France: Organisation for Economic Co- operation and Development.

Pandey, A. (Ed.). (2009). *Handbook of plant-based biofuels*. Boca Raton, FL: CRC Press.

Popp, J. (2009). Economic balance on competition for arable land between food and bioindustry. Retrieved 11 June 2012 from http://www.oecd.org/dataoecd/27/57/42607457.pdf

Searchinger, T., Heimlich, R., Houghton, R. A., Dong, F., Elobeid, A., Fabiosa, J., Tokgoz, S., Hayes, D., and Yu, T. (2008). Use of US croplands for biofuels increases greenhouse gases through emissions from land-use change. *Science*, Vol 319, pp1238–1240.

Siebert, S., Portmann, F. T., and Döll, P. (2010). Global patterns of cropland intensity. *Remote sensing*, Vol 2, pp1625–1643.

Spectrum. (2013). *Wheat: World supply and demand summary*. Retrieved 8 April 2012 from http://www.spectrumcommodities.com/education/commodity/statistics/wheat.html

Steenblik, R. (2007). *Subsidies: The distorted economics of biofuels*. The Global Subsidies Initiative. Geneva: International Institute for Sustainable Development. Retrieved 11 June 2012 from http://www.iisd.org/gsi/sites/default/files/oecdbiofuels.pdf

Stephenson, A. L., Dennis, J. S., and Scott, S. A. (2008). Improving the sustainability of the production of biodiesel from oilseed rape in the UK. *Process Safety and Environment Protection*, Vol 86, pp427–440.

Sugar beet feedstock use increases in Germany. (2013). *Ethanol Producer Magazine*. Retrieved 5 February 2013 from http://ethanolproducer.com/articles/9503/sugar-beet-feedstock-use-increases-in-germany

Te Buck, S., Neeft, J., Smit, A. B., Janssens, S.R.M., Conijn, J. G., Jager, J. H., Prins, H., and Luesink, H. H. (2010). *Greenhouse gas emissions from cultivation of maize, rapeseed, sugar beet and wheat for biofuels*. NUTS-2 report from the Netherlands. Den Haag, the Netherlands: Agentschap NL.

Webb, J., Watson, P., Bellamy, P., and Garstang, J. (2010). *Regional emissions from biofuels cultivation*. Didcot, UK: AEA Energy and Environment.

9 Sugar beet ethanol in the EU

K. W. Jaggard and B. Townsend

9.1 Introduction

Of the many species of plants that accumulate carbohydrates during growth, there are very few that use simple, easily fermentable sugar as their principal energy storage compound. Three species are economically important worldwide: *Saccharum officinarum* (sugarcane), *Beta vulgaris* (sugar beet and fodder beet), and *Sorghum bicolor* (sweet sorghum). Of these, *B. vulgaris* is the only one that can be grown successfully outside tropical regions.

Beets have been grown as a field crop only since the seventeenth century, when they were grown in Europe for cattle fodder. A range of *B. vulgaris* types of various colours were grown either for their foliage or for the storage root. A white storage-root type was the parent material used to breed the modern sugar beet. Today, the plants are biennial, harvested while still vegetative, and eventually achieve a sucrose concentration in the storage root that can exceed 20% of the fresh weight. At present, the most commonly grown types are bred for sugar extraction (beets are the source of about 20% of the world's sucrose consumption), but there are variants that are used for cattle fodder. At harvest time, these variants usually have higher water contents than sugar beets and larger concentrations of chemical components that would inhibit the extraction and purification of sucrose in a beet sugar factory. These compounds need not be inhibitors of biofuel production.

Apart from the sugar, the beet root consists mostly of cell-wall material – approximately one-third each of cellulose, hemicelluloses, and pectins – and very little lignin. This co-product, beet pulp, is usually used as a sugar-rich and fibre-rich animal feed that is particularly suitable for ruminants and monogastrics (Harland et al., 2006). Increasingly, today it is being used for the production of biogas, but in the future it may be used as a feedstock for cellulosic ethanol fuel.

At present, beet and beet products are used for ethanol fuel production in Belgium, the Czech Republic, France, Germany, and the UK (Figure 9.1). Until the area expanded rapidly from about 2005, almost all production was in France. Now approximately 100,000 ha are grown for bioethanol production, about 80% of which is in France and Germany and about 10%

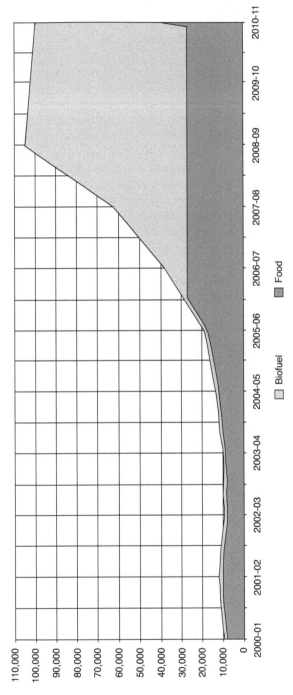

Figure 9.1 Recent changes in the area (ha) of sugar beet used for bioethanol production in Europe

Source: CIBE/CEFS (2010).

in England. In these countries, the bioethanol area is approximately 10% of the total sugar beet area. Ethanol production for the potable market is not included in these data. In describing beet production in Europe, we have confined ourselves to data from these countries.

9.2 Sugar beets in the EU

Beets have been grown in Europe as a crop for cattle fodder since the seventeenth century. At that time, cane sugar was prohibitively expensive because it was imported from the Orient or the Caribbean and then had to be refined. Honey was not available to everyone, and most people had to be content with fruits and berries as their only sources of sweetener. Therefore, there was considerable interest to find an alternative to the sugar that came from cane. In 1747 the German chemist Marggraf reported to the Prussian Academy of Science that he had obtained sweet-tasting crystals from an alcoholic extraction from the roots of beets, and that the crystals were of the same composition as the crystals obtained by refining the juice of sugarcane (Francis, 2006). However, he was able to crystallize no more than about 1.6% of the root's fresh weight, so consequently little was done to exploit Marggraf's discovery for the next four decades. Then, one of Marggraf's students, Franz Carl Achard, obtained patronage from the Prussian royal family and built the world's first beet sugar factory in Lower Silesia. In 1802 the factory processed about 250 tonnes of beets grown in the previous year and extracted about 4% sugar on a root fresh weight basis. The beet sugar industry had begun.

This industry obtained a huge boost due to the actions of Napoleon the First, Emperor of France. He banned the import of all British goods into Europe in 1806, and in retaliation, Britain blockaded France and cut its links with its colonies. Consequently, the price of cane sugar in continental Europe increased enormously: this provided the conditions for an expansion of beet sugar production. In 1811 about 40 beet sugar factories were established, mostly in northern France, and in 1812 Napoleon decreed that 100,000 ha of beets should be planted in the French empire. The British removed the continental blockade in 1813, the sugar price collapsed, and the majority of the factories closed.

The diminished industry struggled on, mostly in Russia and in France, where sufficient improvements in extraction technology were made for the industry to continue to be profitable, especially because imported sugar was subject to a tariff. Improved technology was sufficient to restart the industry in Germany and in the Austro-Hungarian empire in the 1840s. In the next three decades, beet sugar industries were established successfully in most European countries, but not in Britain and Portugal, where there was a continual conflict of interest between the production of sugar at home and the import of cane sugar from their colonies. It was not until World War I demonstrated the dangers of relying on imports for staple foods that Britain developed a serious interest in homegrown sugar from beets.

The sugar factories throughout Europe were able to continue to operate because they provided the incentive for farmers and seed producers to improve the yield and the quality of the raw material, the beet. Essentially, beets for the biofuel industry today have many requirements in common with beets for sugar extraction: large yield per hectare, high sugar concentration, and low production costs per tonne of fermentable material. All the European sites for the production of biofuel from beets are, or were until recently, sites for sugar extraction.

The use of sugar beets as a feedstock for bioethanol production in Europe stems from two EU policies – the RED and the reform of the Sugar Regime – and from a ruling of the WTO. The RED sets targets for each national government to run its transport fleet with 10% of the fuel coming from renewable (mostly ethanol or biodiesel) sources by 2020. Probably more important was the reform of the Sugar Regime, which cut annual EU sugar production from approximately 18 million tonne to 13 million tonne, thus releasing much sugar extraction capacity to be scrapped or used for other purposes.

The ruling of the WTO outlawed the export of high-cost EU sugar into the world market. This meant that sugar producers had to find profitable ways to deal with occasional production surpluses over and above carrying stocks forward from one year to the next. One way was to convert surplus sugar into ethanol for fuel. In reality, sugar is not refined first; instead, the ethanol is either fermented from a watery beet extract or from molasses. These policies and rulings are all recent, so the production of fuel ethanol has grown from an industry that did not exist less than a decade ago. Because beet-based biofuel is mostly produced in the sugar industry, beets for biofuel use similar genotypes and the same agronomy as beets for sugar extraction. Where ethanol is produced in a sugar factory that operates as a biorefinery, it is usually impossible to distinguish beets for ethanol from beets for sugar.

9.3 Feedstock requirements

Sugar beets are not a major ethanol feedstock in the EU, but their relevance is expected to increase (Table 9.1). While 10 million tonnes were used in 2010, consumption is projected to reach 29 million tonnes in 2020. This will be roughly one-third of the production (up from 9% in 2010).

Table 9.1 Sugar beet use for ethanol production in the EU

	Unit	2000	2010	2020
Use in biofuels	million tonne	Negligible	9.8	29.1
Share of all sugar beets	%	Negligible	9%	28%
Biofuel feedstock harvested area	million ha	Negligible	0.1	0.4

Sources: Calculated from FAPRI-ISU (2011); FAOSTAT (2010–2013).

9.4 Cropping systems and agronomy

Cropping systems and associated agronomy can have large impacts on the amount of energy used to produce a crop and, therefore, can also have a large influence on its value for use as a fuel.

In Europe beets grow throughout all of June, July, and August (when evapotranspiration is high), so the crop is ideally sown into soils that have a large reservoir of water that can be used if rainfall fails to meet the evaporative demand. Beets are therefore most commonly grown in deep, silty loams. There are a few exceptions where the crop is grown in sandy soils (i.e., in eastern England and in parts of Belgium and the Netherlands); in these conditions, the crops either suffer water stress regularly or are irrigated.

Where beets are grown for bioethanol, they are grown in rotation with other crop species. Historically there were serious pest and disease problems that could only be controlled by having a beet-free period of two or more years. However, as growers come to rely more on varieties that are resistant to soil-borne pests and diseases, beet-free periods of only one or two years are becoming more common (Table 9.2), but this practice may not be sustainable. Commonly, beets are grown in rotation with crops of winter and spring cereals, maize, oil seed rape, peas, and potatoes. The common crops immediately preceding beets in France and England are shown in Table 9.2. About 90% of these cereals are sown in autumn.

The objective of beet production is to produce a storage root that contains the maximum amount of sugar and the minimum amount of protein. Consequently, the nitrogen fertilizer application for beet production is very different from that of a cereal crop, where protein yield has often been an important objective. Typically, the beet crop in Europe takes up about 220 kg N/ha, and the harvested beet contains about 100 kg/ha. The dose of fertilizer applied is typically between 100 and 120 kg/ha, but not all of it is mineral nitrogen; much of the beet crop receives organic manure. Today the average mineral nitrogen dose, for example, in southern Germany is 105 kg/ha and in England is 93 kg/ha. In these countries, in common with all other

Table 9.2 Cropping history of the beet fields in France in 2010 and in England in 2011; values are percentages of the cropped area

	Previous Crop		Beet-free Period		
	France	England	Years	France	England
Wheat	70	66	1	10	2
Barley		27	2	18	12
Potato	26	2	3	30	25
Other	4	5	>3	42	61

Sources: Anon. (2011a); R. Limb, pers com.

European beet producers, nitrogen usage in the crop has declined greatly in the past two or three decades, while the sugar yield has been increasing (Figures 9.2–9.4). Sugar beets now have the lowest nitrogen usage of any major arable crop in the UK (Anon., 2011b).

The use of phosphate and potash across Europe is not as well documented as the use of nitrogen, although Draycott et al. (1997) reported that the phosphate and potash use in England and Germany was similar, while the amounts applied in France were much larger and amounts in Belgium were intermediate (Table 9.3). In the past, phosphate and potassium tended to be applied at standard rates, irrespective of soil fertility; consequently, soil nutrient concentrations tended to rise. In general, current application rates are based on analyses of the soil sampled during autumn before beets are grown; therefore, application rates have been declining recently. For example, in England the average amounts of phosphate (P_2O_5) and potash (K_2O) fertilizers applied to beet crops are 20 and 75 kg/ha, respectively, and these amounts have reduced by 73% and 54% since the early 1980s.

Pesticides are essential inputs for efficient beet production. Throughout the countries that use beets for biofuel, the amounts of pesticides being applied have been declining, partly in response to the availability

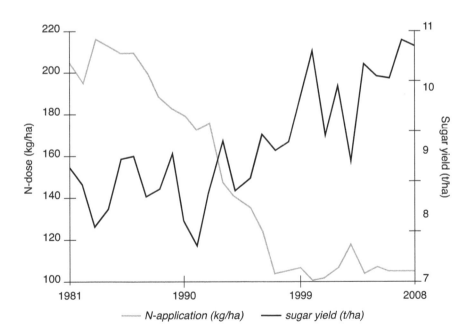

Figure 9.2 Changes in nitrogen fertiliser application rates and sugar yields of beet in southern Germany

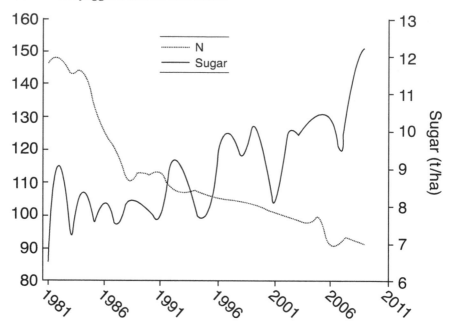

Figure 9.3 Changes in sugar yields and nitrogen fertilizer application rates to beet crops in England

Source: Anon. (2011b).

of chemicals that require smaller doses to achieve the desired effect, and partly in response to improved application techniques. In Belgium, France, Germany and the UK, the average amount of plant protection products applied to beets is now about 4 kg/ha of active ingredient, most of which is herbicide. In all cases, the amounts have been declining while sugar yields have been increasing, so it is now common for usage to be as small as 400 g of active ingredient per tonne of sugar (CIBE/CEFS, 2010). The amounts used per ha on beets are similar to the amounts used on wheat and barley and much less than the amount used on potatoes; an example for the UK in 2008 is given in Table 9.4.

A practice that has experienced increased uptake and, in some cases, increased energy consumption has been the use of soil conservation techniques. To reduce the risk of erosion in spring before the beet canopy is large enough to protect the soil surface from erosion by wind or water, farmers increasingly sow the beet into a mulch (a residue left by the previous crop or from a specially sown cover crop). These cover crops are typically grown

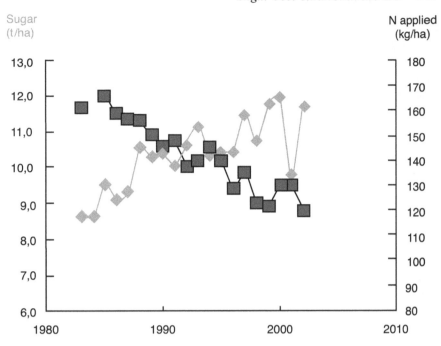

Figure 9.4 Changes in nitrogen fertilizer application rates (squares) and sugar yields (diamonds) of beets in France

Source: After Cariolle and Duval (2006).

Table 9.3 Amounts of nitrogen, phosphate, and potash fertilizers (kg/ha) applied to beet crops in various EU countries in 1995

	N	P_2O_5	K_2O
Belgium	165	126	178
England	105	65	125
France	130	110	240
Germany	110	60	150

Source: After Draycott et al. (1997).

over the winter or during spring to protect the soil; they can also serve to trap nitrogenous minerals that would otherwise leach out of the soil, or to protect the crop seedlings from the activity of cyst-forming nematodes. In Germany, more than 40% of the beet crop is now sown into some form of mulch, a fourfold increase in 20 years. In erosion-prone regions in France, intercrops are now grown on 40–60% of the beet area, compared with 20–30% about 10 years ago.

Table 9.4 Pesticide usage on various crops in Great Britain 2008

Crop Type	Application (kg active ingredient/ha)
Beet	3.92
Winter barley	3.74
Winter wheat	4.98
Oilseed rape	3.21
Potatoes	16.31
Forage maize	1.14

Source: DEFRA pesticide usage reports 224 and 232.

Energy is a major input in beet crop production. In the mid-1990s, Stephan and Kromer (1997) estimated that the crop, from soil preparation to the completion of harvest, consumed about 21.5 GJ/ha. A study in France estimated the energy use as 15.6 GJ/ha, but did not make an allowance for the energy required to manufacture the machinery (Bignon and Cariolle, 1997). Tzilivakis et al. (2005a) estimated the energy needed to grow beets and deliver them to the factory using 12 production systems that represented beet growing in England. On average they estimated that energy consumption, including the energy required to manufacture the machinery, was 21.7 GJ/ha. Mortimer et al. (2004) estimated the energy required to grow a beet crop and deliver it to the factory in England at 22.1 GJ/ha. The estimates made by Hülsbergen and Kalk (2001) for beets in Germany are similar, once a deduction has been made for the energy required for the production of livestock manure, which is used in some cropping regimes. In Europe, wheat is the other crop that is likely to be used for ethanol production. Mortimer et al. (2004) estimated the energy required to produce wheat in the UK at 21.1 GJ/ha, very similar to the 20.8 GJ/ha quoted by Tzilivakis et al. (2005b). Kuesters and Lammel (1999) estimated that 17.5 GJ/ha was required to produce wheat in Germany, but they excluded the energy cost for transporting the grain off the farm.

The energy output–input ratio for beet production for a range of farming systems in the UK was estimated at between 7 and 15 and averaged almost 10 (Tzilivakis et al., 2005b). Keusters and Lammel (1999) estimated energy efficiency values ranging from 11 to 29 for beet production and from 6 to 13 for wheat. However, their yield figures were taken from field experiments, where the most productive areas tend to yield more than they would if production was on a field scale.

Average yields of sugar beets have been rising rapidly in recent years (Table 9.5). In the biofuel-producing countries, the average increase was almost 4% per year during the past five years. Some of this is attributable

Table 9.5 Average sugar yield (t/ha) of beet, as delivered to the factories, in various countries in 2004–2005 and 2009–2010

	2004–2005	*2009–2010*
Belgium	11.1	13.2
Czech Republic	8.1	9.3
France	12.5	14.6
Germany	9.8	11.6
UK	10.4	12.6
EU25	8.4	11.0

Source: ITBFR, http://www.itbfr.org.

to a combination of improvements in the varieties that are sown and in the agronomy, but a significant proportion of the improvement is likely to be the result of a warmer climate (Jaggard et al., 2007). Global warming is likely to result in the continued improvement of beet yields so long as the crops are grown on the most water-retentive soils (Richter et al., 2006): this is unlikely to be the case for cereals, because they do not show a large positive response to being grown in warmer conditions. Beets intended for biofuel production are usually produced on a contract, which specifies a lower beet price than that applied to beets for sugar production, although the beets are indistinguishable. These contracts are usually fully subscribed, so they must be competitively priced in relation to value of the competing crops. Competing crops are those that can act as a break in wheat production (i.e., oilseed rape and maize).

9.5 Conversion to biofuel

An overview of the conversion of beets to ethanol is given in Figure 9.5. At present, all biofuel produced from sugar beets in Europe uses fermentation of the sucrose to produce ethanol; there is no attempt to use the beet pulp (mostly cellulose and hemicellulose). In France some of this ethanol is used as the feedstock to make ETBE, a fuel additive.

The conversion coefficient that is usually quoted is 100 litres of ethanol per tonne of beet, or approximately 7,000 litres per ha at today's average yield. Thus, in the EU where 100,000 ha are devoted to bioethanol, the annual production of ethanol from beet feedstock is approximately 550,000 tonnes. Key figures on beet ethanol production are presented in Table 9.6.

The global warming potential (GWP) of beet and ethanol production is expressed in carbon dioxide equivalents to account for the variable warming potential of the various gasses once they reach the atmosphere; the major gasses in this respect are carbon dioxide, nitrous oxide, and methane. Tzilivakis et al. (2005b) estimated that the GWP ranged from 0.015 to 0.031

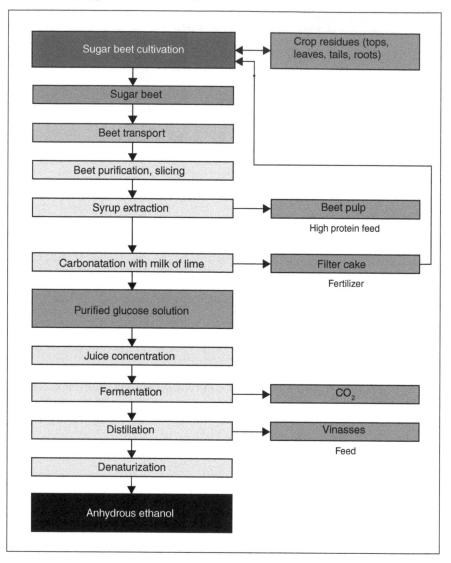

Figure 9.5 Conversion from sugar beet to ethanol. Figure design by H. Croezen.

tonne CO_2 eq. per tonne of beet and averaged 0.024 tonne/tonne for cultivation scenarios in England. In terms of the energy stored in the beet biomass, the GWP averaged 0.0062 tonne CO_2 eq. per GJ of output.

   Probably the most detailed study available of ethanol production from sugar beets was conducted by Mortimer et al. (2004) on behalf of and with the help of British Sugar. The authors had access to the details of the raw materials used to process the beets, their quantities, and the prices. They made life-cycle analysis assumptions and also assumed that the main

Table 9.6 Efficiency of sugar beet-to-ethanol conversion in the EU

	Unit	BioGrace	Straights
Conversion efficiency	litre/tonne	104	98–110
Conversion efficiency	GJ/GJ	0.54	
Biofuel yield	litre/ha	7,200	7,900
Biofuel yield	GJ/ha	153	168
Co-product yield	tonne/ha		0.4

Sources: Calculated from BioGrace (2012); and Straights, http://www.fwi.co.uk/gr/fwistraights.pdf.

co-product was pulp dried for animal feed; other co-products were soil, stones, and lime for agricultural use. Their initial model used a natural-gas-fired boiler and grid electricity to supply energy: the energy requirement was 0.828 MJ/MJ of bioethanol output, considerably more than estimated from other studies (Elsayed et al., 2003; Vignali, 2008).

Mortimer et al. (2004) attributed the difference to their use of more detailed data. They went on to analyse the process using a range of more efficient energy sources and a store of thick beet juice that can be available for fermentation all year. The analysis demonstrated the huge improvements that can be made by using an efficient power source (a gas turbine) and exporting the surplus power to the electricity grid. The energy requirement was cut to 0.36 MJ/MJ and the GWP to 22 g CO_2 eq/MJ (Table 9.7). A gas turbine is the power generator used at British Sugar's Wissington factory, where it produces bioethanol. There, heat is repeatedly recycled to produce a large range of co-products (Anon., 2011b).

Mortimer et al. (2004) also assessed a range of scenarios for producing ethanol from wheat grain. Where the power source was a gas turbine, the process was almost as efficient as using sugar beets. In Brazil, neither the wheat nor the beet systems seem as effective as the use of sugarcane to produce ethanol. Cane processing is fuelled by burning the bagasse (the fibrous

Table 9.7 Summary of results for bioethanol production from sugar beet and wheat grain

	Energy Requirement	Carbon Dioxide Output	Methane Output	Nitrous Oxide Output	GWP
	(MJ/MJ)	(g CO_2/MJ)	(g CH_4/MJ)	(g N_2O/MJ)	(g CO_2 eq/MJ)
Beet, natural gas, grid	0.828	41	0.074	0.011	47
Beet, CHP, gas turbine	0.360	19	0.007	0.010	22
Wheat CHP, gas turbine	0.404	21	0.009	0.039	33

Note: CHP is Combined Heat and Power.

Source: Mortimer et al. (2004).

part of the cane stalk). The GWP of cane-based ethanol production is estimated at about 20 g CO_2 eq/MJ (Vignali, 2008), very similar to the value for
sugar beets based on a gas turbine system. However, the nitrous oxide component of all these systems is still the subject of large uncertainties, and the
value currently associated with sugarcane is small. Wheat has an advantage
because the grain does not easily perish so long as it is stored carefully, so it
is available as a substrate year-round. Sugar beets respire in storage and thus
lose sugar; even in cool conditions they will lose about 0.14% of their sugar
per day (Jaggard et al., 1997). Therefore, to provide a continuous supply of
raw material for fermentation, it is necessary to extract the sugars into a juice
during the autumn and winter months so that it can be stored for later use.

These conversion pathways were also modelled for beet production in the
Netherlands by Corré and Langeveld (2008) using the Energy Crop Simulation Model (E-CROP) and beet yields of 74 tonnes per ha. This yield equates
to about 12.5 tonnes per ha of sugar and is close to the average being produced in northwestern European countries where beet ethanol is currently
manufactured (Table 9.5). Corré and Langeveld simulated production of
ethanol and animal feed from the pulp, ethanol plus methane from the pulp,
and ethanol and methane from the beet plus the foliage. In practice, it would
be difficult to use both the pulp and the tops for energy production without
hauling large quantities of wet material over long distances. However, the
ethanol plus animal feed and ethanol plus pulp biogas chains are realistic on
an industrial scale, and their analysis shows that large amounts of renewable
energy can be produced from this source while reducing GHG emissions
substantially (Table 9.8). Corré and Langeveld also considered ethanol production from sugarcane in Brazil and concluded that beet and cane sources
produced similar energy output per ha and similar reductions in GHG emissions per MJ of energy fixed.

The well-to-wheels study (Vignali, 2008) calculated the GHG balance of
the ethanol from beet (pulp to fodder) pathways and estimated that about

Table 9.8 Modelled output from use of a beet crop to produce ethanol or ethanol
plus biogas

	Unit	Ethanol + Animal Feed	Ethanol + Biogas from Pulp	Ethanol + Biogas from Pulp and Tops	Ethanol from Sugarcane
Ethanol	tonne/ha	5.5	5.5	5.5	5.5
Methane	tonne/ha	0.0	0.7	1.8	0.0
Gross energy output	GJ/ha	176.0	217.0	276.0	165.0
Net energy output	GJ/ha	93.0	123.0	170.0	141.0
Net GHG reduction	tonne CO_2-eq/ha	6.1	6.9	10.3	8.9

Source: Corré and Langeveld (2008).

60 g CO_2 eq. would be saved for every MJ of energy output: this estimate is very similar to that of Corré and Langeveld (2008). The output–input ratios of the processing of beets to ethanol depend hugely upon the mix of products. The well-to-wheels study (Vignali, 2008) and the Mortimer et al. (2004) study both assumed that the main products were ethanol and beet pulp for animal feed. However, important considerations are the size of the markets for the co-products and the way that the co-products must be presented to meet the needs of the market. In the case of beet pulp for fodder, fresh pulp is a low-value product that must be used quickly. Its economic value is greatly increased if it is conserved by being dried and pelleted, but this drying process is energy intensive.

Also, the animal feed market is of finite size, and biofuel production from wheat will add greatly to the amount of feed that is available, thus squeezing the opportunities to sell dried beet pulp. These changes may mean that beet pulp is used in anaerobic digesters to produce biogas in order to fuel part of the ethanol and sugar production process instead of being used for fodder; this is being investigated by some European sugar producers and is expected to save about 70% of the energy needed for ethanol manufacture (Vignali, 2008). To date, none of the published studies have included two other important co-products: CO_2 and betaine. Carbon dioxide is produced during fermentation, and in these studies this is considered as being vented to the atmosphere; instead, it can be collected, purified, and compressed for other applications. Betaine is extracted from beet juice and is used as feed for fish and as a feedstock in the pharmaceuticals industry.

The importance of the value and use of the co-products was again emphasised by Rettenmaier et al. (2008). They calculated life-cycle outcomes for a range of bioenergy pathways, including bioethanol from beets or wheat. Their values for net energy output and GHG reduction were 87 GJ/ha and 7.8 tonne CO_2 eq/ha for an ethanol and beet pulp production system, very similar to the values in Table 9.7. However, they went on to take into account the fodder production that would be displaced by the pressed pulp: it was assumed that the displaced fodder was barley grain. Rettenmaier and colleagues assessed the land area not planted with barley and instead assumed it was planted with beets for biofuel, which would again produce pulp that would replace yet more barley. The result was that the GHG reduction became 33 tonne CO_2 eq. for every ha of land originally devoted to biofuel beets. This saving is huge, but again it depends greatly on finding a market for all the pressed pulp. If the pulp has to be dried to meet the market requirements, then the savings are much less.

9.6 Discussion

At present, the market for sugar produced in Europe is constrained by national quotas so that surplus production in one year has to either be held in stock to form part of the next season's quota, or it has to find a non-food use that is free of quotas. One such use, for some of the sugar producers, is biofuel ethanol. European sugar production has decreased since 2006, and in some countries beets for bioethanol are being grown in place of beets for sugar. Therefore, there have been no direct land-use changes as a consequence of this ethanol production. Indirect land use change (ILUC) depends upon the crop that would be grown in place of beets. The animal feed component of the beet crop could be produced most cheaply by growing feed barley, but the amounts of fodder produced per unit area are approximately the same, so there are no ILUCs that need to be considered. On average, one ha of beets for sugar or biofuel will produce about four tonnes of pulp (85% d.m.). This is in contrast to cane-based bioethanol: cane does not have a valuable co-product, and expansion of the cane area in Brazil displaces grazing, which in turn may displace the Cerrado or virgin rainforest, potentially incurring large carbon debts (Fargione et al., 2008).

The use of sugar from sugar beets to produce ethanol has a long history, but for most of that history it was potable ethanol. Ethanol for fuel is less valuable and so must be made more cheaply. In Europe it also has to compete with ethanol imported from North or South America, where it might have been produced from maize or sugarcane. Therefore, the profitability and popularity of beet production on the arable farms in Europe are crucial to the future success of beet-for-biofuel processors. Despite the revision of the EU Sugar Regime and the reductions in beet and sugar prices since 2006, beets have remained popular and therefore profitable with many northwest European farmers.

In most cases farmers get paid less for beets that are used to produce biofuel than for beets used to produce sugar, so the profitability of beets to the farmer will depend upon the relative prices and the relative tonnages of these two product streams. In the bioethanol-producing countries, little more than about 10% of the beet is devoted to ethanol, and few individual farmers will commit more than about 20% of their total beet production to ethanol. In these conditions it is the beet-for-sugar that must carry the fixed costs of the enterprise; beets for bioethanol are still a marginal activity. On the farm, beets have to compete with other break crops for their place in the rotation, and they must also compete with wheat grown continuously. So far, the sugar producers have offered farmers sufficient financial incentives for beets to compete successfully with these alternative crops.

One aspect of the beet's ability to compete profitably against other crops is the likely yield in the future. Throughout the past decade, the yields of beets in northwestern Europe have been increasing at between 1% and 2% per year. Recently this rate has accelerated (Table 9.5). Recent increases in yield in the main competitor crops in European rotations are smaller than that for beets. In addition, beet yield is very sensitive to the rate at which the foliage canopy expands in spring, and this is driven by the temperature (Jaggard and Qi, 2006). Climate change is likely to make European springs warmer, which will increase beet yield still further; the competing crops are much less likely to benefit from this shift (Jaggard et al., 2007).

As with anything else made from beets, an important drawback is the fact that the raw material is perishable. It is only available for the harvest period plus a storage period. In most bioethanol-producing areas of Europe, this availability extends from mid-September to mid-January, or about 120 days. The milder winter weather in England allows this to be extended to about 160 days. The short operating periods mean that investment in employees and machinery seems expensive. The apparent inefficiency in the use of resources is overcome in some factories by extracting juice from the beet during autumn and winter and storing a portion of this for use during the remainder of the year. This has a huge impact on the efficiency of biofuel production from beets because the fermentation and distillation equipment can be operated all year.

Biofuel production

Since the implementation of the biofuel policy, the EU biofuel industry has developed rapidly, with biodiesel production earning an especially important position. However, production still is lagging behind consumption. Ethanol is imported mainly from Brazil, while biodiesel imports vary. High feedstock costs remain a major obstacle for the domestic industry, while investment subsidies are less than those reported for Brazil and the USA. A public debate on sustainability performance of biofuels and a long-awaited policy adjustment (addressing the issue of ILUC) have added to an atmosphere of uncertainty.

Changes in land use

Under current conditions, EU sugar beet cultivation for ethanol is extremely small. While almost 9% of the beets are used in biofuel production, land area involved remains low at 0.1 million ha in 2010. Projected areas for 2020 will occupy less than 0.5% of all arable land, while generating up to a quarter of all domestically produced ethanol. This makes beets an interesting feedstock.

Production efficiency

Under current cultivation conditions in the EU, sugar beets are a highly productive crop that can generate more ethanol per ha than does sugarcane under average Brazilian management. Average yields amount to 7,900 litres of ethanol or 168 GJ per ha (Table 9.9). In addition, large amounts of pulp are generated, and this makes excellent animal feed. Nitrogen productivity is modest, but the high conversion efficiency (as compared to cereals) provides high energy yields per kg of nitrogen fertiliser applied; phosphorus response is among the highest (slightly below that of sugarcane). However, the crop does not have a positive impact on soil organic matter, and beet leaves, which are high in nitrogen, pose a risk for nutrient leaching when not removed after harvest.

Table 9.9 Performance and impacts of sugar beet biofuel chains in the EU

	Unit	Sugar Beet
Crop yield	tonne/ha	79.1
Biofuel yield	l/ha	7,900
Biofuel yield	GJ/ha	168
Co-product yield	tonne/ha	4.0
Nitrogen productivity (crop)	kg dm/kg N	165
Water requirement (biofuel)	m^3/GJ	23
Nutrient productivity (biofuel)	GJ/kg N	2.3
	GJ/kg P_2O_5	1.8
Risk of nutrient leaching	–	High
Impact on soil organic matter	–	Negative

In many parts of the world, the water footprint of a process is at least as important as the carbon footprint. Where water is scarce, it is important to use it efficiently, and agriculture is the largest consumer of water. In northern Europe, where the atmosphere during the growing season is usually warm (not hot) and moist, beet crops consume approximately 400 mm of water from sowing until harvest. During this period, water consumption is equivalent to about 250 m^3 per tonne of sugar stored in the beet (Dunham, 1993). This is equivalent to approximately 42 m^3 per 100 litres of ethanol. Allowing for the energy density of ethanol, this equates to a water footprint of 17.5 m^3/GJ of gross energy output. However, this calculation is based on the performance of crops in experiments: real crops appear to be less productive.

In European beet-production terms, it is fair to assume that the yield gap between experimental crops and commercial beet production will be about 25% (Jaggard et al., 2010), so the water footprint increases to 23 m^3/GJ. This takes into account the water used for beet growth: processors of the crop are usually net exporters of water (Anon., 2011b), so they add nothing to the water footprint. Gerbens-Leenes et al. (2009) showed that use of sugar beets for ethanol production had a smaller water footprint than any other crop. As with all other calculations of the efficiency of production chains, it is important to take into account the value of co-products, and that has not been included in these water footprint estimates.

GHG emissions

Following RED and EU regulations, beet cultivation and conversion would generate some 40 g CO_2 eq/MJ of ethanol plus an ILUC reference value of 12 g CO_2 eq/MJ. Total emissions would amount to 52 g, which is equivalent to a GHG emission reduction level of 38% (Table 9.10). Assuming average emission levels as suggested by Mortimer et al. (2004) and Laborde et al.

Table 9.10 GHG emissions of EU sugar beet ethanol chains

	Unit	*BioGrace*	*Laborde*
Biofuel production	g CO_2-eq/MJ	40.3	22–47
ILUC	g CO_2-eq/MJ	12.0	7
Total	g CO_2-eq/MJ	52.3	35.0
Reduction	%	38%	58%

Sources: BioGrace (2012); Mortimer et al. (2004); Laborde et al. (2011).

(2011), and disregarding ILUC emissions, total emissions would be no more than 35 g CO_2 eq/MJ. In this case, GHG reduction would be about 60%. These findings are confirmed by Corré and Langeveld (2008), who showed that GHG reduction can be increased by two-thirds through using co-products in biogas production to support distillation energy requirements.

Economic and social impacts

Sugar beet cultivation and conversion into sugar have been playing an important role in European agriculture, and both the crop and the plants have earned their position in the agricultural landscape – physically (as in farm and rural areas) and economically. The crop has long been considered one of the backbones of arable farm incomes in many regions, especially where rainfall is high enough to guarantee sufficient availability of water for its cultivation.

Biomass availability

Production of ethanol from sugar beets is not expected to affect the crop's availability for food markets. Crop yields in the EU are very high, and yields still are being improved. This makes the sugar beet one of the most efficient biofuel feedstocks, as confirmed by very large per-ha (sugar and) ethanol production levels. Sugar beets can be well integrated in existing crop rotations, although their share should remain limited due to their susceptibility to soil-borne pests and diseases. They generate considerable quantities of co-products.

9.7 Conclusion

The use of sugar beets as a feedstock for the production of bioethanol is efficient in terms of energy yield, energy recovery, and savings of GHG production. In these terms it is at least as efficient as first generation bioethanol from wheat. If these systems are powered by a gas turbine that supplies heat and power, then they are as efficient as Brazilian cane-based ethanol, and in direct land-use terms, both have similar potential to reduce

GHG production. When the indirect land use is considered, bioethanol from beets has a considerable advantage because its co-product (pulp for animal feed) yield is equivalent to the yield of barley grown in the same area. However, any new beet ethanol production factory would be very capital intensive. Therefore, in Europe, ethanol from beets is exclusively produced at what are, or were, sugar production sites where most of the capital has already been invested and where additional capacity can be made available cheaply.

References

Anon. (2011a). *Rapport d'activité: 2010–2011.* Institut Technique de la Betterave: Paris.

Anon. (2011b). *UK beet sugar industry sustainability report 2011.* Retrieved from www.nfusugar.com/ . . . /UK-Beet-Sugar-Industry-Sustainability-Report-2011/

Bignon, J., and Cariolle, M. (1997). Analyse du cycle de vie de l'ETBE, hypothèses culturales et industrielles, conséquences et principaux résultants. In *Proceedings of the 60th IIRB Congress,* pp 141–158. Cambridge, UK.

BioGrace. (2012). Retrieved 12 May 2012 from http://www.biograce.net/content/ghgcalculationtools/calculationtool

Cariolle, M., and Duval, R. (2006). Nutrition: Nitrogen. In A.P. Draycott (Ed.), *Sugar beet.* Oxford, UK: Blackwell Publishing.

CIBE/CEFS. (2010). *The EU beet and sugar sector: A model of environmental sustainability.* Retrieved from www.cibe-europe.eu

Corré, W.J., and Langeveld, J.W.A. (2008). *Ethanol from sugarbeet in the Netherlands: Promising biofuel yields and GHG emission reduction.* Retrieved from www.biomassresearch.eu/images/FactsandFigures_Sbeet.pdf

Draycott, A.P., Allison, M.F., and Armstrong, M.J. (1997). Changes in fertilizer usage in sugar beet production. In *Proceedings of the 60th IIRB Congress,* pp 39–54. Cambridge, UK.

Dunham, R.J. (1993). Water use and irrigation. In D.A. Cooke and R.K. Scott (Eds.), *The sugar beet crop.* London: Chapman & Hall.

Elsayed, M.A., Matthews, R., and Mortimer N.D. (2003). *Carbon and energy balances for a range of biofuels options.* Project number B/B6/00784/REP. URN 03/836. Sheffield Hallam University, Sheffield, UK.

Fargione, J., Hill, J., Tilman, D., Polasky, S., and Hawthorne, P. (2008). Land clearing and the biofuel carbon debt. *Science,* Vol 319, pp1235–1238.

Francis, S.A. (2006). Development of sugar beet. In A.P. Draycott (Ed.), *Sugar beet.* Oxford, UK: Blackwell Publishing.

Gerbens-Leenes, W., Hoekstra, A.J., and van der Meer, T.H. (2009). The water footprint of bioenergy. *Proceedings of the National Academy of Sciences,* Vol 106, pp10219–10223. doi/10.1073/pnas.0812619106

Harland, J.I., Jones, C.K., and Hufford, C. (2006). Co-products. In A.P. Draycott (Ed.), *Sugar beet.* Oxford, UK: Blackwell Publishing.

Hülsbergen, K.J., and Kalk, W.D. (2001). Energy balances in different agricultural systems: Can they be improved? *Proceedings of the International Fertilizer Society,* No. 476.

Jaggard, K. W., Clark, C.J.A., May, M.J., McCullagh, S., and Draycott, A. P. (1997). Changes in the weight and quality of sugarbeet (*Beta vulgaris*) roots in storage clamps on farms. *Journal of Agricultural Science*, Vol 129, pp287–301.

Jaggard, K. W., and Qi, A. (2006). Agronomy. In A. P. Draycott (Ed.), *Sugar beet*. Oxford, UK: Blackwell Publishing.

Jaggard, K. W., Qi, A., and Ober, E. S. (2010). Possible changes to arable crop yields by 2050. *Philosophical Transactions of the Royal Society, Series B*, Vol 363, pp2835–2851.

Jaggard, K. W., Qi, A., and Semenov, M. A. (2007). The impact of climate change on sugarbeet yield in the UK: 1976–2004. *Journal of Agricultural Science*, Vol 145, pp367–375.

Kuesters, J., and Lammel, J. (1999). Investigations of the energy efficiency of the production of winter wheat and sugar beet in Europe. *European Journal of Agronomy*, Vol 11, pp35–43.

Mortimer, N. D., Elsayed M. A., and Horne R. E. (2004). *Energy and greenhouse gas emissions for bioethanol production from wheat grain and sugar beet*. Final report for British Sugar plc, Resources Research Unit, School of Environment and Development, Sheffield Hallam University, Sheffield, UK.

Rettenmaier, N., Reinhardt, G., Gärtner, S., and Münch, J. (2008). *Bioenergy from grain and sugar beet: Energy and greenhouse gas balances*. Report of the Institute for Energy and Environmental Research: Heidelberg, Germany.

Richter, G.M., Qi, A., Semenov, M.A., and Jaggard, K.W. (2006). Modelling the variability of UK sugar beet yields under climate change and husbandry adaptations. *Soil Use and Management*, Vol 22, pp39–47.

Stephan, C., and Kromer, K-H. (1997). Energiebilanz aktueller Zuckerrubenproduktionstechniken. *Proceedings of the 60th IIRB Congress*, 353–358. Cambridge, UK.

Tzilivakis, J., Jaggard, K. W., Lewis, K. A., May, M., and Warner, D. J (2005a). Environmental impact and economic assessment for UK sugar beet production systems. *Agriculture, Ecosystems and Environment*, Vol 107, pp341–358.

Tzilivakis, J., Warner, D.J., May, M., Lewis, K.A., and Jaggard, K.W. (2005b). An assessment of the energy inputs and greenhouse gas emissions in sugar beet (*Beta vulgaris*) production in the UK. *Agricultural Systems*, Vol 85, pp101–109.

Vignali, M. (2008). Biofuels versus diesel and gasoline in the JEC-WTW report version 2c: An Extract from the *Well-to-Wheels Analysis of Future Automotive Fuels and Powertrains in the European Context*. EUR 23549 EN-2008.

10 Oil palm biodiesel in the Far East

J.W.A. Langeveld, P.M.F. Quist-Wessel and H. Croezen

10.1 Introduction

Palm oil is the world's favourite vegetable oil since its consumption surpassed that of soybean oil in 2005. It is used as a feedstock for the manufacturing of cooking oils, frying fats, margarine, and a wide range of other food applications. Due to its favourable composition, smell, and consistency palm oil is an ideal ingredient in fats, soaps, sauces, powdered milk, and ice cream. It is considered one of the best oils for frying, and as such is used frequently in households, restaurants, and industries around the world. Palm kernel oil and palm kernel meal are by-products of palm oil production. The former has many of the same qualities as coconut oil and is sought for applications in condensed milk, biscuits, and bread; giving these products volume and soft texture and making them last longer. Palm kernel meal is used in the production of foods and as a supplement in animal feed (Agricultural Forum, 2006).

Palm oil increasingly is used as feedstock for biodiesel production, although volumes remain much smaller than those for other major oil crops (soybean, rapeseed). The main producers of palm oil are Indonesia and Malaysia. This chapter discusses the production of palm oil in these countries, and for conciseness, we refer to them as the Far East. Normally, more countries in the region are included in this term, but they do not produce much palm oil.

Indonesia, a large Southeast Asian multi-island nation of 240 million people, is a rapidly evolving middle-income country with agriculture still providing incomes for the majority of the labour force. Agricultural development is closely related to poverty reduction and strengthening of food availability for the poor. The current Medium-Term Development Plan focusses on reducing poverty, improving the quality of the public sector, fostering democracy, and strengthening the rule of law. The longer-term objective is to promote more inclusive growth (ADB, 2012a). Most Indonesian farmers own only a small piece of land (average land ownership is below 0.35 ha), with poverty being prevalent among small-scale farmers and the landless. While more than 32 million Indonesians live below the poverty line, the country's outlook for economic development is good due to a low debt-to-GDP ratio and budget deficit (World Bank, 2013).

Malaysia, a country with 29 million inhabitants that lies to the northeast of Indonesia, is a middle-income country with a recent record of strong economic performance and poverty reduction. Over the past few years, strong progress was made in poverty reduction and education. Malaysia is on track to meet goals related to infant mortality and birth attendance. Economic growth, 7.2% in 2010, is supported by a strong export and rising domestic demand, particularly investment (ADB, 2012b). Malaysia covers some 33 million ha and has a strong background in the production of industrial crops, especially palm oil production.

While both countries have many similarities (e.g., soils, climate, a large forest area, and strong population growth), differences exist in land area, size of the population (Indonesia being much larger), role of subsistence agriculture in the economy (larger in Indonesia), and economic growth (stronger in Malaysia). In terms of biofuel policy and palm oil production, the situation in both countries is comparable. Differences will be stipulated where applicable.

The remainder of this chapter is organised as follows. First, the chapter analyses land cover and land use (Section 10.2). Next, it discusses the development of a biofuel policy and industry (Section 10.3). In Section 10.4, feedstock requirements for domestic biofuel production are defined while Sections 10.5 and 10.6 elaborate on palm oil production and its conversion to biodiesel, respectively. This is followed by an evaluation (Section 10.7) and some conclusions (Section 10.8).

10.2 Land resources

Indonesia and Malaysia have a tropical humid climate. Together, they cover some 214 million ha of land, mostly forest (54%). Figure 10.1 depicts a schematic overview of different land cover types. One-third of the land is agricultural area, the remaining 17% is classified as 'other' (neither agriculture nor forest).

Indonesia covers 181 million hectares (ha), half of which is forest area. One-third is agricultural, which is dominated by cereals (rice) and oil crops (coconut, palm oil) (data retrieved from FAOSTAT between 2010 and 2013, http://faostat.fao.org). Sawah (lowland) rice area covered some 8 million ha in 2000, with 4.5 million ha of irrigated area, rain-fed land, and (tidal) swamp areas (ADB, 2006). Palm (oil) and cocoa area increased by more than 10% annually in the 1980s and early 1990s, reaching 4 million ha in 1999 as a result of strong world market demand. Of the 10 million ha of estate crop area expansion, 7 million ha were developed by smallholders; private companies and state-owned enterprises developed the remainder (ADB, 2006). Malaysia, covering almost 33 million ha, consists of two-thirds forest. Agriculture occupies 19% of the land, of which two-thirds (4 million ha) is used for oil palm.

Together, the two countries host some 62 million ha of agricultural land (Table 10.1), which consists mainly of arable land (25 million ha) and agricultural tree crops (26 million ha) including food and industrial crops like

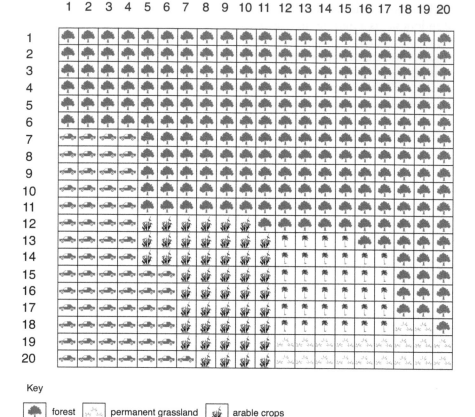

Figure 10.1 Schematic overview of land use in the Far East

No geographical representation. Each cell represents 0.25% of land area. Position of the categories was chosen randomly.

Source: Calculated from FAOSTAT (2010–2013), http://faostat.fao.org

coffee, tea, oil palm, coconut, and rubber. Grassland area is relatively small. There is quite a lot of 'other' land (37 million ha on a total land area of 214 million ha), suggesting that this not only comprises urban areas but also abandoned agricultural land. Data on fodder crops could not be obtained. Fallow land is around 11 million ha or one-third of the arable area, which seems rather high. Cropping intensity is high due to widely practiced double and triple cropping, especially on irrigated lowlands.

The past has seen a continuous increase of agricultural land use (arable crops, tree crops) and 'other' land (Table 10.1). The amount of harvested crop area per unit of arable land (MCI) increased by nearly 30%, mostly in the past two decades. A large amount of forest (25 million ha) has been lost. The average annual loss of forest area amounted to 2 million ha in the 1990s but

Table 10.1 Land area and land use in Indonesia and Malaysia

	Unit	1980	1990	2000	2010
Forest area	million ha	No data	141	121	115
Other land	million ha	No data	21	39	37
Agricultural area	million ha	43	45	54	62
Permanent grassland	million ha	12	13	11	11
Agricultural tree crops	million ha	8	12	20	26
Arable area	million ha	18	20	22	25
Of which:					
arable crop area	million ha	20	22	27	31
fodder crop area	million ha	No data	No data	No data	No data
fallow area	million ha	No data	No data	9.3^1	11.3^2
Multiple Cropping Index (MCI)	–		1.02	1.04	1.15	1.21

[1]Figure for 2001.
[2]Figure for 2005.

Source: Calculated from FAOSTAT (2010–2013), http://faostat.fao.org.

declined after 2000. Agricultural tree crop area (including oil palm) has more than doubled since 1990.

10.3 Markets and policy conditions

Biofuel policies have not been very ambitious. A 5% blending obligation for 2025 was installed in Indonesia in 2006 (Table 10.2). One year later, a subsidy of about $100 million was announced for farmers growing biofuel crops. Blending rates for biodiesel were halved in 2008 when palm oil prices were very high. Recently, they were raised to 7.5%. As of 2012 companies in the coal and mining industries have to obtain 2% of their energy requirements from biofuels.

Malaysia implemented its first biofuel policy in 2006 (Table 10.2), followed by a Biofuel Industry Act, which introduced a 5% biodiesel blending

Table 10.2 Biofuel policies in Indonesia and Malaysia

Year	Indonesia	Malaysia
2006	Presidential instruction No. 1/2006 National Policy (5% blend by 2025)	National Biofuel Policy Malaysian Biofuel Industry Act
2007	Interest rate subsidy of Rp 1 trillion (100 million $) for farmers growing biofuel crops	
2008	Reduction of blend Biosolar (B5) and Biopertamax (E5) from 5% to 2.5%	5% biodiesel blend from palm oil and methyl ester
2012	Biodiesel blend 7.5%; coal and mining to consume 2% biofuels	

Sources: APEC (2012); Chin (2011).

Table 10.3 Biodiesel market development in Indonesia and Malaysia

	Unit	Indonesia 2007	Indonesia 2010	Malaysia 2007	Malaysia 2010
Biodiesel production	billion litre/year	0.01	0.1	0.1	0.1
Biodiesel consumption	billion litre/year	0.0	0.0	0.0	0.02
Support investments	billion US$	No data	No data	No data	No data
Subsidies, tax exemptions	US$/litre	0.0	0.31[1]	0.0	No data
Import levies	US$/litre	No data	No data	No data	No data
Production cost	US$/litre	0.0	No data	0.0	0.54[2]

Note: Billion = thousand million.
[1] As of 2013.
[2] Data for 2006.
Sources: Calculated from FAPRI-ISU (2011); Hamelinck et al. (2011).

obligation in 2008. This obligation requires 0.6 billion litres of biodiesel. Implementation of the act later was delayed due to soaring palm oil prices. The government is waiting for prices to reach a modest level before reinstalling a mandate.

Biodiesel production is very modest, reaching some 230 million litres in 2010. Consumption is very low (Table 10.3). For 2020, FAPRI-ISU projects a production of 1 billion litres and a consumption of only 140,000 litres. High palm oil prices have raised biodiesel production costs (APEC, 2012). Biofuels in Indonesia are subsidised at the same level as fossil fuels. As of 2013 subsidies for biodiesel are 3,000 Rupiah (US$0.31) per litre. Subsidies for ethanol are 3,500 Rupiah (US$0.36). All in all, perspectives for biodiesel production remain bleak.

10.4 Feedstock requirements

In the period 1995–2010 global palm oil production more than tripled to an annual output of 47 million tonnes. The main producers are Indonesia (47%) and Malaysia (38%). Indonesia has surpassed Malaysia as the leading palm oil producer. Thailand and the Andes region in Central America have also shown a growing importance as palm oil producers. In the next decade production of palm oil is expected to increase considerably due to demographic and economic developments (MVO, 2009).

Palm oil production in Indonesia and Malaysia doubled between 1990 and 2000, and has doubled again since then (Table 10.4). The area devoted to oil palm has shown similar increases, rising to 9 million ha in 2010. Yields declined in the 1980s but have improved since 1990. Average annual

Table 10.4 Palm oil production in Indonesia and Malaysia

	Unit	1980	1990	2000	2010
Harvested area	million ha	1	2	5	9
Production	million tonne of FFB	16	42	93	178
Yield (three-year average)	tonne FFB/ha	17.7	17.4	18.5	19.2
Average annual increase[1]	kg FFB/ha/year		−29	146	142

[1]Figure for 1980–1990 presented in column 1990; figure for 1990–2000 in 2000, and so on.

Sources: Calculated from FAOSTAT (2010–2013), http://faostat.fao.org; FAPRI-ISU (2011).

improvement is modest, with about 146 kg per ha per year. Oil palm yields in Indonesia strongly increased in the 1980s, followed by a decline in the 1990s and a recovery in the 2000s. The situation is less favourable in Malaysia, where yield improvement is declining.

Palm oil in Malaysia is mostly produced on estates owned by plantation companies that often operate mills where oil is extracted. Small-scale farmers with varying sizes of oil palm holdings normally sell fruit bunches to dealers who send them to nearby mills (Basiron, 2007). Oil palm trees are relatively old (25–30% older than 20 years) and show declining yields; replanting has been delayed due to high palm oil prices. Lack of skilled labour is further reducing harvest activity. The growing labour problem may be the largest issue facing Malaysian estates, which rely heavy on immigrant (Indonesian) workers. Yields are further depressed by adverse weather conditions (e.g., El Niño, which causes drought, and La Niña, which produces heavy rains) and declining fertiliser use (USDA FAS, 2011).

Malaysia has limited opportunities to expand palm production. Palm oil producers are looking across land borders for options to expand plantations, particularly in Indonesia where land and labour still are available. According to FAPRI, future expansion in Indonesia will be mostly generated by an enlargement of the oil palm area to seven million ha (+22%).

The use of oil palm for biodiesel so far has been modest. Consumption amounted to 520,000 tonnes in 2000 and close to 1 million tonnes in 2010 (Table 10.5), involving less than 1% of palm oil. This may increase to 2% in 2020, when feedstock production may reach 4.4 million tonnes. Land use so far has been negligible.

Table 10.5 Oil palm use for biodiesel production in Indonesia and Malaysia

	Unit	2000	2010	2020
Use in biodiesels	million tonne	0.5	1.0	4.4
Share of all palm oil	%	0.6%	0.5%	1.8%
Biodiesel feedstock area[1]	million ha	0.03	0.05	0.21

[1]Assuming yield improvement will be similar to that in the period 2000–2010.

Source: Calculated from FAPRI-ISU (2011).

10.5 Crop cultivation

Oil palm (*Elaeis guinensis*), native to West Africa, is the highest yielding oil crop in the world. It produces palm oil (extracted from the flesh – *mesocarp* – of fresh fruits) as well as palm kernel oil (from the seed kernel). The commercial crop comes in three main varieties, which vary in fruit composition and oil contents. Oil palms may live up to 200 years, but commercial yields decline rapidly after 30 years (Verheye, 2012).

Cultivation starts with removal of the vegetation, which may be primary or old plantation crops (oil palm, rubber, cocoa). Trees may be removed or piled for natural decomposition but often are burnt (Musim Mas, 2007), although this is not always allowed. Seedlings are raised in nurseries before being planted in rows at a distance of 9 m at a density of some 150 plants per ha. Triangular planting allows maximum penetration of sunlight (Basiron, 2007).

Ground cover is quickly established by planting cover crops to prevent soil erosion and weed growth. Oil palm starts bearing from the fourth year onwards, and its economic life varies from 30 to 35 years. Under average management conditions, a mature plantation (8–9 years old) may yield 15–18 tonnes of fresh fruit bunches (FFB) per ha. Under good maintenance and management, yield may reach 25–30 tonnes (Agricultural Forum, 2006). Harvesting is done using chisels or hooked knives attached to long poles (Basiron, 2007; El Bassam, 2010). Fruit bunches, wedged in the leaf axils, contain 1,000–4,000 fruits. They weigh 15–25 kg, occasionally reaching 50 kg. Fruits make up 50–65% of the weight (Verheye, 2012).

A typical crop calendar is presented in Figure 10.2. In Indonesia the best period for planting is June–December (Agricultural Forum, 2006). Humid climates allow year-round harvesting, but in regions with distinct dry and rainy periods, flowering mainly takes place in the dry periods and harvest takes place in the wet seasons (El Bassam, 2010). Time from flowering to harvesting is 5–6 months (Verheye, 2012). Harvesting rounds in Malaysia are organised throughout the year so that the same palm is visited every two weeks (Basiron, 2007).

Soils for oil palm should be deep and well drained. The best pH range is from 5.5 to 7, although the palm will thrive at lower pH values when receiving sufficient fertilisers (El Bassam, 2010). Soils should have high inherent fertility because oil palm is more nutrient demanding than other palm crops.

Activity	Jan	Feb	Mar	Apr	May	Jun	Jul	Aug	Sep	Oct	Nov	Dec
Planting						X	X	X	X	X	X	X
Growing season	X	X	X	X	X	X	X	X	X	X	X	X
Fertilizing					X	X			X	X		
Harvesting	X	X	X	X	X	X	X	X	X	X	X	X

Figure 10.2 Oil palm cropping calendar in Indonesia/Malaysia

Ideal soils are deep, medium loam soils that are rich in humus (Agricultural Forum, 2006), but new plantations are often located on peat soils. Draining of peat soils will lead to its decomposition, which releases nutrients and high amounts of carbon. Decomposition can be limited by maintaining high groundwater tables (MPOB, 2011).

Initially, palms can be intercropped with food crops that are replaced by a cover crop to suppress weeds. Legume cover crops provide valuable nitrogen (El Bassam, 2010). Data on input use are presented in Table 10.6. Typical fertiliser application amounts to 200 kg of nitrogen, 115 kg of phosphate, and 145 kg of potassium oxide per ha. Both higher (Hardon et al., 2001) and lower applications (BioGrace, 2012) have been reported for nitrogen.. Data refer to mineral soils, as decomposition of peat soils may generate large amounts of nutrients, thus reducing the need for fertiliser applications. Fertilisers should preferably be applied in three to four split doses (Agricultural Forum, 2006). Animal manure applications are not common, but application of green leaf manure or compost is advantageous, especially where the soils are poor in organic matter.

Residues from palm oil production may be used as fertiliser to reduce the need for artificial fertilisers. Palm fibre, empty fruit bunches (EFB), and kernel shell are used as solid fuel for the mills while EFB are also applied as mulch. Treated palm oil mill effluent (POME) is another source of nutrients (Musim Mas, 2007). POME is generated by combining the following fractions from the extraction process (Corley and Tinker, 2003):

- Condensate from bunch sterilization (0.6 t/t palm oil produced)
- Water phase or sludge from oil clarification centrifuges (up to 2.5 t/t palm oilproduced)
- Water from the hydro cyclone used in separation (0.25 t/t palm oil)

Table 10.6 Oil palm input use in Indonesia and Malaysia

Parameter (unit)	Per ha	Source
Nutrient application (kg/ha)		
N	200	Typical (Agricultural Forum, 2006)
	430	High yielding crop (Hardon et al., 2001)
	128	BioGrace (2012)
P_2O_5	115	Typical (Agricultural Forum, 2006)
	144	BioGrace (2012)
K_2O	145	Typical (Agricultural Forum, 2006)
	200	BioGrace (2012)
Application of agro-chemicals	4.9	Pleanjai and Gheewala (2009)
(kg active ingredient/ha)	8.4	BioGrace (2012)
Diesel (litre/ha)	63	BioGrace (2012)
Required rainfall (mm)	1,800–2,000	Hardon et al. (2001)

In other words, the primary source of liquid effluent or POME is the clarification process, although the amount produced per ton FFB varies greatly, depending on the process used. Hardon et al. (2001) reported gross nutrient uptake of high yielding palms on a marine clay at 200 kg of nitrogen, 65 kg P_2O_5, and 310 kg of K_2O per ha. In practice, uptake may be lower. Some 30–40% of the nutrients are removed by the FFB, 25–35% is returned to the soil as dead leaves and male inflorescences, and the rest is immobilised in the trunk (Hardon et al., 2001).

According to Basiron (2007), oil palm is little affected by pests and diseases. It requires modest applications of agro-chemicals. Plantations in Malaysia routinely implement biological pest control measures; for example, using owls to reduce rodent populations. According to Verheye (2012), however, oil palm is sensitive to many pests and diseases. Main diseases include spear and bud rot, leaf spot disease, and dry basal rot. Damage of insect pests (e.g., *Ocytes* spp. and *Ronochophorus* spp.) can be controlled mostly by sanitary measures, but occasionally, insecticides will be needed (El Bassam, 2010). Using data provided by Pleanjai and Gheewala (2009), we calculated herbicide applications in Thailand at 5 kg of active ingredient per ha; BioGrace reports even higher applications. Diesel use is modest at 63 litres per ha.

Oil palm requires a well-distributed rainfall and a temperature range of 19–33°C. Because it is a fast-growing crop with high productivity and biomass production, it may require adequate irrigation. The crop responds well to drip irrigation. Mature palms require a minimum of 150 litres per day. Older trees may require up to 200 litres (Agricultural Forum, 2006).

10.6 Palm oil to biodiesel

FFB must be milled within 24 hours after harvest to preserve oil quality. They are sterilized while separating fruitlets from the bunches. Fruitlets then are mechanically pressed to obtain crude palm oil (CPO) and palm kernel oil. Waste and water are removed, followed by refining, bleaching, and deodorizing (RBD). The resulting oil may be separated (fractionated) into liquid and solid fractions by controlled cooling, crystallization, and filtering. The liquid fraction (olein) can be used as a cooking oil; to produce specialty fats, glycerol, and fatty acids (MVO, 2009); or to produce biodiesel.

Oil production and conversion into biodiesel is presented in Figure 10.3. Refined oil is submitted to esterification where it reacts with light alcohols (typically methanol or ethanol) using bases as catalysts. Ethanol is a preferred alcohol given its environmental features (Pandey, 2009).

Conversion efficiency of palm oil biodiesel is some 230 litres of biodiesel per tonne of FFB (Table 10.7). Palm oil gives the highest biodiesel yield per ha (4,500 litres) of all major oil crops. The main co-product is palm meal, which is produced in considerable amounts. Some improvement in conversion efficiency has been projected (FAPRI-ISU, 2011).

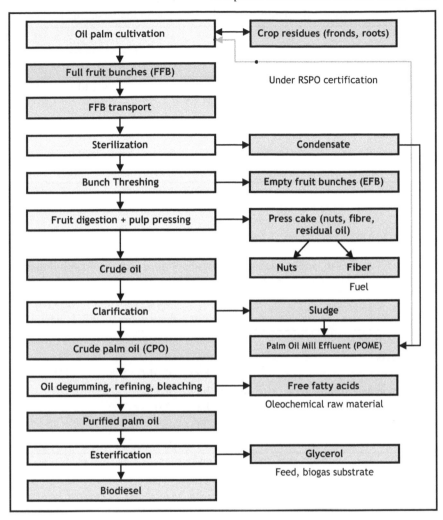

Figure 10.3 Conversion from palm oil to biodiesel. Figure design by H. Croezen and F. Quist-Wessel.

10.7 Evaluation

Biodiesel production

Indonesia and Malaysia are major palm oil producers, but so far domestic (ethanol and) biodiesel production been very limited. The main reason for low biodiesel production is the rise in palm oil prices, which coincided with implementation of the biofuel policy. High production costs have limited the competitiveness for biofuel production. This seems to be more the case for Malaysia where palm oil production is hampered by restrictions in land availability, land shortages, and poor oil yields.

Table 10.7 Efficiency of oil palm-to-biodiesel conversion in Indonesia and Malaysia

	Unit	BioGrace	Other Sources		
			2000	*2010*	*2020*
Conversion efficiency	litre/tonne			223–230	239
Conversion efficiency	GJ/GJ	0.50			
Biodiesel yield	litre/ha	4,500	4,200	4,200–4,600	
Biodiesel yield	GJ/ha	149			
Press cake yield	tonne/ha			4.0–14.0	
Glycerol yield	tonne/ha			0.5	

Sources: Calculated from Croezen et al. (2013); FAPRI-ISU (2011); Van der Vossen and Umali (2001); BioGrace (2012); Galeon (2013).

Table 10.8 Changes in land availability for food and feed production in Indonesia and Malaysia 2000–2010

	Unit	Impact on Land Availability	Share of Land
Increase in biodiesel production	million ha	−0.022	100%
Co-products	million ha	0.010	45%
Losses total	million ha	−0.012	55%
Change in agricultural area	million ha	8.9	
Generated by increased MCI	million ha	2.0	
Land gains total	million ha	10.9	
Net land balance		10.9	

Source: Calculated from Tables 10.1, 10.5, and FAOSTAT (2010–2013), http://faostat.fao.org.

Changes in land use

Land availability has not been affected by domestic biofuel production. Net loss of arable area (estimated at 12,000 ha in 2010) is negligible. In fact, arable crop production expanded with some three million ha while increases in MCI generated another 2.0 million ha of harvested crops (Table 10.8).

Expectations for the near future are favourable. Total oil palm biomass needed is expected to grow to some 4.4 million tonnes, which is expected to cover only 0.2 million ha. This will offer no threat to the production of food and feed crops in the Far East, nor to the export of palm oil or any other product.

Existing concerns for future palm oil production in Malaysia are not related to domestic biodiesel production, but they might be associated with the use of palm oil in biofuel production elsewhere. To our knowledge, however, this can hardly be the case (palm oil consumption in biodiesel production in the EU amounted to no more than two million tonnes in 2010, which represents only 5% of total production in the Far East). While there will be other countries that use imported palm oil in their biofuel industry, this does not necessarily have to be obtained from the Far East.

This does not mean that the concerns for Malaysia listed previously are not important. Declining yields and expected land shortages can affect the palm oil industry in this country, and this may lead to an increased pressure on forest land in the entire region.

Production efficiency

Palm oil biodiesel performance is depicted in Table 10.9. Crop yield is high with 18.4 tonnes per ha, realising a biodiesel production of some 4,200 litres per ha. More than four tonnes of press cake is produced. A leguminous crop yielding two tonnes per ha is assumed to cover the soil. About 35% of fertilisers are removed with the main product, and 30% are recovered from crop residues (leaves, male florescence). The remainder is stored in the crop (all following Hardon et al., 2001). Nitrogen output is 61 kg of dry matter per kg of nitrogen fertiliser. Output per kg of phosphate is rather low. Soil organic matter balance is slightly positive.

Table 10.9 also provides information for palm oil biodiesel produced on weathering peat soils. Crop yield is not expected to be different, but as mineralisation of the peat is generating large amounts of nitrogen and – to a lesser extent – phosphorus, fertiliser applications can be strongly reduced. Consequently, output per kg applied is raised. Impact for soil organic matter, however, is disastrous, and large amounts of carbon will be released.

Any forest clearing for palm oil production should be strongly discouraged. We could not identify any direct relation with domestic biodiesel production. Indirect effects may result from land-use changes for biofuels elsewhere. Possible indirect effects are discussed in Chapter 15.

Table 10.9 Performance and impacts of oil palm biodiesel chains in Indonesia and Malaysia

	Unit	*Oil Palm (clay)*	*Oil Palm (peat soil)*
Crop yield	tonne FFB/ha	18.4	18.4
Biofuel yield	litre/ha	4,200	4,200
Biofuel yield	GJ/ha	90	90
Co-product yield	tonne/ha	4.2	4.2
Nitrogen productivity (crop)	kg dm/kg N	61	500
Water requirement (biofuel)	m^3/GJ	200	200
Nutrient productivity (biofuel)	GJ/kg N	0.5	9.9
	GJ/kg P_2O_5	0.9	1.7
Risk of nutrient leaching	–	Modest	High

Environmental impacts of oil palm production depend strongly on the type of vegetation that is replaced, as well as the type of soil on which it is cultivated. The most negative effects have been reported for removal of humid forest vegetations on peat soils.

GHG emissions

Oil palm is a highly productive crop with an output-to-input energy ratio of 9:1 compared to 3:1 for other oilseed crops such as soybean or rapeseed (Basiron, 2007). GHG emissions related to palm oil biodiesel in the Far East are presented in Table 10.10. Emissions occurring during fuel production (crop cultivation, transport to the plant, conversion) are quite high at 58–68 g of CO_2-eq/MJ, generating a GHG emission reduction of 24–36%. Emissions associated with ILUC are 24–55 g CO_2-eq/MJ, which basically brings total emissions above that of fossil diesel. Installing oil palm purely on former fallow area would, however, allow a small net GHG reduction (calculated from Darlington et al., 2013).

Economic and social impacts

While oil palm production plays an important role in providing employment and generating income in Malaysia, its relevance is much smaller in Indonesia where population is greater and food production is much larger. A typical estate of 2,000 ha of oil palm would employ a staff of nine field workers, a manager, and three assistant managers plus a larger number of manual workers to carry out weeding, fertiliser application, and harvesting. Moreover, the estate sector has attracted a growing number of smallholders to venture into oil palm cultivation. The Malaysian government created special agencies to consolidate or aggregate small areas into estates of an economic size and to provide management and infrastructural inputs (Basiron, 2007).

Because biofuel production in Indonesia and Malaysia so far has been limited, its impact on employment or national economies will be negligible.

Table 10.10 GHG emissions of oil palm biodiesel chains in Indonesia and Malaysia

	Unit	*Results*
Biodiesel production	g CO_2-eq/MJ	58–68
Reduction[1]	%	24–36%
ILUC	g CO_2-eq/MJ	24–55
Total	g CO_2-eq/MJ	82–124

Note: Biofuels for domestic use.
[1]ILUC related emissions not included. Calculated with a fossil reference of 90 g CO_2-eq / MJ.

Sources: Calculated from BioGrace (2012); European Union (2009); Laborde (2010); Darlington et al. (2013); CARB (2012).

Table 10.11 Changes in food biomass availability in Indonesia and Malaysia 2000–2010

	Unit	Impact on Biomass Availability	Share of Biomass
Increase total biomass production	million tonne	48.0	
Biomass in biodiesel production	million tonne	−0.5	100%
Generation of co-products	million tonne	0.3	24%
Net impact of biodiesel	million tonne	−0.1	76%
Net change in biomass availability	million tonne	47.7	
Increase in BP	million tonne	27.9	

Notes: Biomass production at three-year average arable crop level.

Source: Calculated from Table 10.5 and FAOSTAT (2010–2013), http://faostat.fao.org.

Some subsidies are allocated, which will have lead to an improvement of local employment perspectives. The impacts will, however, remain small in comparison to those of palm oil production focussing on domestic or export markets. Furthermore, actual impacts will depend on the way production is organised. This holds especially for the way smallholders are involved in the management of the estates that convert their fresh bunches and the way that rights of previous landowners or users have been guaranteed when land is converted.

Changes in biomass availability

Introduction of a biodiesel policy did not affect production of food, feed, or fibre crops. While only one million tonnes of FFB has been allocated to biodiesel production, a quarter of this is recovered in by-products (mainly palm meal). The net impact on (local) biomass availability is negligible. Arable crop area has expanded (providing 42 million tonnes of extra biomass in 2010) while the output of biomass per ha of arable land has increased by 30% since 2000, generating over 40 million tonnes extra biomass (Table 10.11). Net biomass availability has increased by 85 million tonnes.

10.8 Conclusion

Indonesia and Malaysia have implemented modest biofuel policies. Biodiesel production is not expected to be harmful for profitable palm oil exports. Domestic production is low, and there seems much room for growth. Throughout the past decade, most increases of crop production were realised by expanding agricultural area. This brings large risks for deforestation and subsequent carbon losses. While the performance of clay soil–based oil palm chains is generally favourable, the opening of peat lands and the lowering of groundwater tables to provide land for palm oil plantations will have detrimental impacts on carbon releases.

References

ADB (Asian Development Bank). (2006). *Strategic vision for Indonesia.* Retrieved 17 January 2013 from http://www.adb.org/sites/default/files/pub/2006/Strategic-Vision-Indonesia.pdf

ADB (Asian Development Bank). (2012a). *Indonesia factsheet.* Retrieved 19 January 2013 from http://www.adb.org/sites/default/files/pub/2012/INO.pdf

ADB (Asian Development Bank). (2012b). *Malaysia factsheet.* Retrieved 19 January 2013 from http://www.adb.org/sites/default/files/pub/2012/MAL.pdf

Agricultural Forum. (2006). Forum: Oil palm cultivation. Retrieved 5 October 2012 from http://www.agricultureinformation.com/forums/questions-answers/12541-oil-palm-cultivation.html

APEC. (2012). Malaysia biofuels activities. Retrieved 12 January 2013 from http://www.biofuels.apec.org/me_malaysia.html

Basiron, Y. (2007). Palm oil production through sustainable plantations. *European Journal of Lipid Science and Technology,* Vol 109, pp289–295. DOI 10.1002/ejlt.200600223.

BioGrace. (2012). Retrieved 12 May 2012 from http://www.biograce.net/content/ghgcalculationtools/calculationtool

CARB. (2012). Carbon intensity lookup table for gasoline and fuels that substitute for gasoline. Retrieved 21 September 2013 from http://www.arb.ca.gov/fuels/lcfs/lu_tables_11282012.pdf

Chin, M. (2011). *Biofuels in Malaysia: An analysis of the legal and institutional framework.* CIFOR working paper 64. Retrieved 17 January 2013 from http://www.cifor.org/publications/pdf_files/WPapers/WP64CIFOR.pdf

Croezen, H.J., Bergsma, G.C., Odegard, I.Y.R., and Langeveld, J.W.A. (2013). *De bodem in de bio-economie* [Soils in the biobased economy]. Delft, the Netherlands: CE Delft.

Darlington, Th., Kahlbaum, D., O'Connor, D., and Mueller, S. (2013). *Land use change greenhouse gas emissions of European biofuel policies utilizing the Global Trade Analysis Project (GTAP) model.* Novi, USA: Air Improvement Resource.

El Bassam, N. (2010). *Handbook of bioenergy crops: A complete reference to species, development and applications.* London: Earthscan.

European Union. (2009). *Directive 2009/28/EC of the European Parliament and of the Council on the promotion of the use of energy from renewable sources amending and subsequently repealing Directives 2001/77/EC and 2003/30/EC.* Brussels, Belgium: European Commission.

FAPRI-ISU. (2011). *World agriculture outlook.* Retrieved 18 May 2012 from http://www.fapri.iastate.edu/outlook/2011/

Galeon. (2013). Retrieved 12 January 2013 from http://www.galeon.com/francisko3/balancebio.pdf

Hamelinck, C., Koper, M., Berndes, G., Englund, O., Diaz-Chavez, R., Kunen, E., and Walden D. (2011). *Biofuels baseline 2008.* Retrieved 10 December 2012 from http://www.ecofys.com/files/files/ecofys_2011_biofuels_baseline(2008).pdf

Hardon, J. J., Rajanaidu N., and van der Vossen, H.A.M. (2001). *Ealais guineensis* Jacq. In H.A.M. van der Vossen and B. E. Umali (Eds.), *Plant resources of South East Asia: No. 14 vegetable oils and fats* (pp. 85–93). Leiden, the Netherlands: Backhuys Publishers.

Laborde, D. (2010). *Assessing the land use change consequences of European biofuel policies: Final report.* Washington, DC: International Food Policy Research Institute.

MPOB (Malaysian Palm Oil Board). (2011). *Best management practices of oil palm cultivation on peat land.* Retrieved 4 October 2012 from http://www.rspo.org/sites/default/files/3-%20MPOB%E2%80%99s%20Guidelines%20for%20Oil%20Palm%20on%20Peat%20(Tarmizi).pdf

Musim Mas. (2007). *The Musim Mas best management practices (BMP) in oil palm cultivation and processing.* Retrieved 5 October 2012 from http://www.musimmas.com/images/PDF/Musim%20Mas%20Best%20Management%20Practices%20(BMPs)23rdJuly2007.pdf

MVO (Product Board for Margarine, Fats and Oils). (2009). *Fact sheet palm oil, product board for margarine, fats and oils.* Rijswijk, the Netherlands. Retrieved 17 January 2013 from http://www.mvo.nl/LinkClick.aspx?fileticket=jsFVMZwZzkc%3d&tabid=2301&mid=3341&language=en-US

Pandey, A. (2009). *Handbook of plant-based biofuels.* Boca Raton, FL: CRC Press.

Pleanjai, S., and Gheewala, S. H. (2009). Full chain energy analysis of biodiesel production from palm oil in Thailand. *Applied Energy,* Vol 86, pp209–214.

USDA FAS. (2011). *Malaysia: Obstacles may reduce future palm oil production growth.* Retrieved 8 January 2013 from http://www.pecad.fas.usda.gov/highlights/2011/06/Malaysia/

Van der Vossen, H.A.M., and Umali, B. E. (Eds.). (2001). *Plant resources of South East Asia: No. 14 vegetable oils and fats.* Wageningen, the Netherlands: Prosea.

Verheye, W. (2012). *Growth and production of oil palm.* EOLSS. Retrieved 5 October 2012 from http://www.eolss.net/sample-chapters/c10/E1–05A-27–00.pdf

World Bank. (2013). Indonesia overview. Retrieved 17 January 2013 from http://www.worldbank.org/en/country/indonesia/overview

11 Biofuel production in southern Africa

*P.M.F. Quist-Wessel, J.W.A. Langeveld,
M. van den Berg and W.J. Leonardo*

11.1 Introduction

Fifteen countries in the southern part of Africa organised themselves into the Southern African Development Community (SADC). Originally established as a coordinating conference (SADCC) in 1980, it was transformed into an intergovernmental organisation in 1992 aiming to promote sustainable and equitable economic growth and socioeconomic development through productive systems, cooperation and integration, good governance, and durable peace and security among member states. The main objective on foreign policy is coordination and cooperation, but a tighter coordination is in progress for trade and economic policy with a view to one day establish a common market with common regulatory institutions (World Bank, 2013b).

The SADC is a region full of contrasts among member states and even within countries. This variation is perhaps more characteristic of the region than any generalised overview provides, even if coherent datasets were available. This chapter focuses on two countries, which together comprise much (albeit it not all, by far) of the variation: Mozambique and South Africa.

Mozambique is one of the fastest-growing economies in Africa with excellent access to southern African regional markets. Natural resources include gas, coal, minerals, and timber while, since the mid-1990s, access to farming land has been given to foreign investors. The country, with a population of 24 million (INE, 2013), boasts one of the highest gross domestic product (GDP) growth rates – since the peace agreement in 1992 – but remains one of the poorest countries in the world (Jones and Tarp, 2012). The percentage of Mozambicans living below the poverty line is estimated at 55%, with a marginal rise in rural poverty (from 55% to 57%) since 2002 (Alfani et al., 2012).

The agricultural sector employs approximately 80% of the Mozambican labour force and accounts for about 30% of its GDP and 4% of total exports. The main farming system in Mozambique is rain-fed subsistence farming, with low levels of productivity. Cassava is the main crop in terms of volume, followed by sugarcane and maize (Leete et al., 2013). The traditional use of bioenergy (fuel wood and charcoal) represents a huge share of total Mozambican energy supply.

South Africa is one of the larger and economically more powerful members of the SADC. Its transition to a constitutional democracy was a strong demonstration of the proposition that a peaceful, negotiated path from conflict to cooperation and reconciliation is possible. South Africa is an active participant in international fora such as the International Monetary Fund, the World Bank, the G-20, and the G-24. It is also a member of the Cairns group of agricultural exporting countries and of the group of major emerging countries commonly referred to as BRICS (Brazil, Russia, India, China and South Africa).

In spite of its economic strength, well-established institutions, two decades of political stability, and undeniable poverty-reducing achievements, the country displays high and persistent levels of inequality, exclusion, and unemployment for an upper middle-income country. It stands as one of the most unequal countries in the world, with very high (25%) unemployment rates and limited access to economic opportunities and basic services for the poor. Development challenges include a low life expectancy, the largest number of people with HIV/AIDS in the world (over 5.5 million), and an epidemic of tuberculosis.

The importance of the agricultural sector in the South African economy is rather modest. It employs 5% of the South African labour force and accounts for 3% of its GDP (World Bank, 2013a). However, there are strong linkages into the economy, so that the agro-industrial sector comprises about 12% of GDP, and about 8.5 million people (16%) depend directly or indirectly on agriculture for their livelihoods (Government of South Africa, 2013). The earlier mentioned inequalities also apply to land ownership, with 55,000 farmers owning 85% of arable land (World Bank, 2013a). The largest area of farmland planted with field crops is maize, followed by wheat, sugarcane, and sunflower. Maize and sugarcane are the main staple crops in terms of volume and value.

The South African economy, closely linked with the world economy, suffered from worsened global economic conditions, in particular the slowdown in Europe and China – the two main export destinations for the country. This has weakened growth prospects and fiscal revenues. GDP growth was projected at 2.5% in 2012, down from 3.1% in 2011. Growth is projected to pick up by 2014, but bottlenecks in electricity supply capacity are likely to form an important constraint for further development.

South Africa's development strategy faces a number of significant challenges, including accelerating growth and sharing its benefits more broadly, developing opportunities to all, and improving delivery of public services. The Medium Term Strategic Framework (MTSF), running from 2009 to 2014, formulated priorities on issues such as inclusive economic growth, sustainable livelihoods, economic and social infrastructure, rural development, food security, land reform, and sustainable resource management and use (World Bank, 2013b).

The remainder of this chapter is organised as follows. First, land cover and land use are analysed (Section 11.2). Next, the chapter discusses the development of biofuels policy and industry (Section 11.3). Section 11.4 explains to what extent domestic crops are involved in biofuel production, while Sections 11.5 and 11.6 elaborate on the production of sugarcane, sweet sorghum and Jatropha and their conversion to biofuels. This is followed by an evaluation (Section 11.7) and some conclusions (Section 11.8).

11.2 Land resources

Mozambique

Statistics for land use show large variation and there appears to be discrepancy between different sources. FAOSTAT was chosen as the main source to ensure some level of coherence but not because it is necessarily of better quality than other sources. According to FAOSTAT, Mozambique covers 80 million ha with 49 million ha of agricultural area (Table 11.1). The agricultural area mainly consists of permanent grassland (89%), including 10% arable land, while tree crops occupy less than 1% (0.2 million ha). A 15% share of fallow is assumed in line with findings in Brazil. Not many dedicated fodder crops are expected; most food crops offer residues that can be used as animal feed. Frequency of harvesting, expressed as the ratio of crop area harvested per unit of arable land (Multiple Cropping Intensity [MCI]) is low for a tropical country, but it is increasing. Since the end of the civil war in the early 1990s, agricultural land use (arable crops, tree crops) has increased while 4 million ha of forest was lost, most likely due to fuel wood

Table 11.1 Land area and land use in Mozambique

	Unit	1980	1990	2000	2010
Forest area	million ha	No data	43	41	39
Other land	million ha	No data	No data	No data	No data
Agricultural area	million ha	47	48	48	49
Permanent grassland	million ha	44	44	44	44
Agricultural tree crops	million ha	0.2	0.2	0.3	0.2
Arable area	million ha	2.9	3.5	3.9	5.2
Of which:					
arable crop area	million ha	2.4	2.9	3.3	4.4
fallow	million ha	0.4	0.5	0.6	0.8
Multiple Cropping Index (MCI)	–		1.08	1.02	0.98	1.03

Source: Calculated from FAOSTAT (2010–2013), http://faostat.fao.org.

Table 11.2 Land area and land use in South Africa

	Unit	1980	1990	2000	2010
Forest area	million ha	No data	9	9	9
Other land	million ha	No data	15	13	15
Agricultural area	million ha	95	97	100	97
Permanent grassland	million ha	81	83	84	84
Agricultural tree crops	million ha	0.8	0.9	1.0	0.4
Arable area	million ha	12.4	13.4	14.8	12.5
Multiple Cropping Index (MCI)	–	0.85	0.74	0.55	0.53

Source: Calculated from FAOSTAT (2010–2013), http://faostat.fao.org.

collection, shifting agriculture, forest fires, timber exports, and lack of plans for land use.

South Africa

The total terrestrial land area of South Africa is 121 million ha, of which 80% (97 million ha) is agricultural land (Table 11.2). About 87% of this is used for extensive grazing, mainly because of rainfall constraints. The remaining 13% is used for the production of arable crops (12.5 million ha) and permanent crops. Only 1.4 million ha (about 10% of total arable land) are under irrigation, but these produce a significant proportion of the country's agricultural output, notably horticultural and vegetable production and viticulture (FAO, 2005). Cropping intensity has declined rapidly. In 1980, on average, nearly 85% of arable land was harvested, however, by 2010 this declined to 53%.

11.3 Markets and policy conditions

According to Janssen and Rutz (2012), South Africa and Mozambique have the most advanced regulatory frameworks for bioenergy in Africa. Table 11.3 presents an overview of biofuel policy development. Mozambique aims to include both large-scale and smallholder farming systems to produce biofuels in order to reduce import dependency, to promote sustainable socioeconomic rural development, and to reduce fuel price volatility. This shall be supported by gradual introduction of biofuel blending with petrol and diesel (Government of Mozambique, 2011). The main driver for policy support to biofuels in South Africa is the need to create a link between the country's first and second economies. The aim is to stimulate economic development and alleviate poverty through promoting farming in areas that were previously neglected (DME, 2007; Letete and Von Blottnitz, 2012).

The biofuel discussion in Mozambique has been prominent since the election campaign in 2004. While interest in Jatropha for biodiesel spearheaded

Table 11.3 Biofuel policies in Mozambique and South Africa

Year	Mozambique	South Africa
1998	Adoption of energy policy	White Paper on Energy Policy
2003		White Paper on Renewable Energy: targeting 4% renewable energy by 2013
2007		Biofuels industrial strategy: 2% blending of biofuels (400 million litres/year) by 2013, representing 30% of the renewable energy target
2009	National Biofuels Policy and Strategy (NBPS)	NERSA's renewable energy feed-in tariff programme
2012	Biofuels policy strategy: a blending regulation (3% in volume of biodiesel; 10% in volume of anhydrous ethanol)	

Sources: Janssen and Rutz (2012); Government of Mozambique (2009, 2011); DME (2003, 2007).

the political promotion of biofuels, there was also significant private sector and government interest in the production of ethanol. The principal feed-stock considered was sugarcane, although interest also began to be shown in sweet sorghum due to its potential multiple purposes: food, feed and fuel. To avoid potential competition between food crops and energy crops, a zoning exercise was conducted in the country. The results indicated the availability of about 7 million ha of land (i.e., 140% of the current arable land area) for commercial agricultural activities (Schut et al., 2010).

The Mozambican government approved a National Biofuel Policy and Strategy in 2009, promoting biofuel production while limiting potential negative impacts. Some of the political and strategic pillars are the following: limits on land allocation based on agroclimatic land zoning, approval of selected feedstocks (sugarcane and sweet sorghum for ethanol; coconut and Jatropha for biodiesel), use of sustainability criteria to select investment projects and allocate land titles, the creation of a domestic biofuel market with blending mandates (gradually phased in at increasing levels from 2012), increase export to generate tax revenues and foreign currency, promotion of regional markets for biofuels, and the establishment of tariffs for the purchase of electricity produced from biomass, particularly cogeneration of electricity from ethanol production (Government of Mozambique, 2011; Schut et al., 2010).

As of 2012 the blending regulation was launched with a compulsory blending of E10 (10% of ethanol with 90% of gasoline) and B3 (3% of biodiesel with 97% of fossil diesel) until 31 December 2015. After this date these levels are to be further increased to E15 and B7.5 until 2020 and to E20 and B10 as from January 2021 (Government of Mozambique, 2011).

South Africa initiated a renewable energy policy by launching a *White Paper on Renewable Energy* in 2003, targeting 4% of renewable energy by 2013. In 2007 the South African government released the National Bio-fuels Industrial Strategy, proposing a biofuels market penetration target of 2% of national liquid fuel supply by 2013, or 400 million litres per year to be based on local agricultural and manufacturing production. The level of biofuel blending was revised down from 4.5% as proposed in the draft strategy document of 2006 to 2%, mainly due to food security concerns (DME, 2007). The 2% target represents about 30% of the national renewable energy target for 2013 (Letete and Von Blottnitz, 2012).

During 2009 the National Energy Regulator of South Africa (NERSA) announced the South Africa renewable energy feed-in tariff programme to further stimulate the renewable energy sector. Crops proposed for biofuel production include sugarcane and sugar beet for ethanol and sunflower, canola, and soybeans for biodiesel. Maize and Jatropha are excluded due to alleged negative impacts on food security (maize) and the potential invasiveness of Jatropha into natural areas (DME, 2007; Janssen and Rutz, 2012). The strategy targets new and additional land and estimates that about 14% of the arable land in South Africa, mainly in the former homelands, is currently underdeveloped, whereas only about 1.4% of arable land will be required to achieve the 2% target (Letete and Von Blottnitz, 2012).

In Mozambique, an emerging biodiesel sector is based on the use of coconut oil and occasionally palm oil as feedstock (Econergy, 2008). Because the prices of coconut oil went up significantly, the opportunity cost of using the oil for biodiesel rather than sale on the international market was too high. Biofuel production is modest at 68 million litres for Mozambique and 63 million litres for South Africa in 2010 (Table 11.4). Consumption figures are at the same level as production figures.

Table 11.4 Biofuel production and market development in southern Africa

	Unit	Mozambique		South Africa	
		2005	2010	2005	2010
Ethanol production	billion litre/year	Neg	0.023[1]	Neg	0.015[2]
Ethanol consumption	billion litre/year	Neg	0.023[1]	Neg	0.015[2]
Biodiesel production	billion litre/year	Neg	0.045[2]	Neg	0.048[2]
Biodiesel consumption	billion litre/year	Neg	Neg	Neg	0.048[2]

Notes: billion = thousand million; Neg = negligible.
[1]Figure for 2009.
[2]Figure for 2007–2009.

Source: Calculated from EIA (2013); OECD-FAO (2009–2010).

Mozambique and South Africa are signatories to several trade agreements that establish the terms and conditions for access to the country's potential biofuel production to key regional and international markets, namely the EU, the USA, and the SADC. Mozambique has duty-free access to the US market under the Generalized System of Preferences (GSP), which grants reduced duty or duty-free access to developing countries. This was extended by the African Growth and Opportunity Act (AGOA) in 2000, a US trade act that significantly enhances US market access for (currently) 39 sub-Saharan African countries, including Mozambique and South Africa (Schut et al., 2010). Mozambique also has duty-free access to the EU, under the Everything but Arms (EBA) initiative, which grants unlimited duty-free access to the world's least developed countries.

Mozambique and South Africa are members of the SADC Trade Protocol, an agreement between 11 SADC members that is aimed at promoting regional trade. Under this agreement, tariffs on intraregional trade of certain goods have been eliminated or substantially reduced. Tariffs on so-called sensitive goods were eliminated by 2012, although Mozambique has until 2015 to comply. When fully implemented, the protocol will give Mozambican (and South African) products duty-free access to a market of more than 200 million people with a GDP of US$275 billion, with reciprocal treatment for the goods from the other members. However, in the case of biofuels, the final size of the regional market, and access to it, will depend on the establishment of harmonised fuel standards and blending mandates or authorisation in the other member countries (Schut et al., 2010).

11.4 Feedstock requirements

Farming in Mozambique involves over 3.8 million farmers the majority of whom (99%) are smallholders (INE, 2011). A small number of commercial farmers cultivate about 60,000 ha and refurbished agro-industrial units grow some 40,000 ha of sugarcane. Maize and cassava are the major staples; other food crops of significance include sorghum, beans, groundnuts, millet, and rice. The use of purchased agricultural inputs (improved seeds, fertilisers, and pesticides) is limited to a small number of modern farm enterprises growing cash crops and vegetables and to out-growers of tobacco and cotton, producing crops on contract. Cereal yields in the smallholder subsistence sector are generally low, and losses in the field and stores are high (FAO, 2010).

The NL Agency (2012) listed a number of biofuel projects approved by the Mozambican government, including two sugarcane and sweet sorghum ethanol projects requiring 33,000 ha of land to produce 220 million litres per year and five Jatropha biodiesel projects requiring 29,500 ha of land. The government promotes the use of bagasse and Jatropha seedcake for generating electricity. Molasses is not used consistently; some is sold locally and some is exported to Europe, but generally market conditions are limited (Econergy, 2008). In general, there is a concentration of biofuel activities in the Beira corridor around Zambezia and along the southern coast between Maputo and Inhambane (Schut et al., 2010).

Sugarcane for ethanol

Mozambique

Sugar cultivation – a traditional industry in Mozambique – was destroyed during the war. Since 1992, when the war ended, production recovered significantly (Table 11.5). Yields declined from 40 tonnes per ha to 13 tonnes per ha in 1990 but have improved since, reaching about 70 tonnes per ha in 2010 (FAOSTAT, data retrieved during 2010–2013, http://faostat.fao.org). As a result of the rehabilitation program, the area planted to sugar increased to more than 40,000 ha in 2010, resulting in the production of 2.7 million tonnes in 2010.

Sugarcane is mostly grown on large mechanised plantations using substantial amounts of inputs, particularly water and fertilisers (Cheeseman, 2004). In 2009 it covered 40,000 ha, of which about 35,000 were irrigated. The African Development Bank and the World Bank are supporting crop expansion, aiming to add an additional 47,000 ha with an emphasis on out-growers (Econergy, 2008). Out-growers deliver to a central processing unit and often obtain most inputs from the central processing mill, while their only input is labour (Bernd et al., 2012).

The country has four major commercial sugarcane companies and two small ethanol plants for domestic consumption (NL Agency, 2012). Data on ethanol feedstocks are presented in Table 11.6. The use of sugarcane in ethanol production is expected to increase from 0.2 million tonnes in 2010 to 0.6 million tonnes in 2020.

Table 11.5 Sugarcane production in Mozambique

	Unit	*1980*	*1990*	*2000*	*2010*
Harvested area	million ha	0.05	0.03	0.03	0.04
Production	million tonne	2.0	0.33	0.39	2.7
Yield (three-year average)	tonne/ha	40	13	16	68
Average annual increase[1]	kg/ha/year		–2,723	307	5,200

[1]Figure for 1980–1990 is presented in the column for 1990; figure for 1990–2000 is presented in the column for 2000, and so on.

Sources: Calculated from FAOSTAT (2010–2013), http://faostat.fao.org; FAPRI-ISU (2011).

Table 11.6 Sugarcane use for ethanol production in Mozambique

	Unit	2000	2010	2020
Use in ethanol	million tonne	Negligible	0.2	0.6
Share of all sugarcane	%	Negligible	7%	No data
Biofuel feedstock area	million ha	Negligible	0.003	No data

Sources: Calculated from OECD-FAO (2010); Burrell (2010).

South Africa

Sugarcane area in South Africa amounts to 430,000 ha, three-quarters of which is suitable for harvest every year. About 20% of the crop is grown under irrigation, mainly in the northern regions. The sugar industry operates in South Africa under the jurisdiction of the Department of Trade and Industry where it has retained a measure of price support and some protection against imports. The industry further distinguishes itself by operating a small-scale out-grower scheme (Cartwright, 2010). There are currently 27,000 registered growers producing some 20 million tonnes of sugarcane per year in areas extending from the Eastern Cape Province through KwaZulu-Natal to Mpumalanga. Large-scale growers are responsible for approximately 83% of the total sugarcane production, while 9% of the total crop is produced by small-scale farmers and 8% by milling companies (Department of Agriculture, Forestry and Fisheries, 2013). Production in 2010 yielded 16 million tonnes of sugarcane (Table 11.7).

South Africa already manufactures reasonable volumes of bioethanol by fermenting molasses from the sugar industry. This bioethanol is not used in fuel but as potable alcohol, in paints and inks, and by the pharmaceutical industry (Table 11.8).

Sweet sorghum for bioethanol

Sorghum (*Sorghum bicolor* [L.] Moench) is the fifth-most cultivated cereal in the world after wheat, rice, maize, and barley. It is mostly grown in the semi-arid tropics where production is constrained by poor soils, low and erratic rainfall, and low input use (Rao et al., 2009). Many African small-scale farmers are familiar with grain varieties that require similar management to sweet sorghum. Sorghum is drought tolerant, has a short growth cycle, and is not very demanding in terms of nutrients and soil quality. Sweet sorghum is a multipurpose crop: stalk juice can be used for ethanol production or sweet syrup, seed for human consumption, and bagasse as cattle feed. Its juice, however, is not suited for the production of crystal sugar.

The short harvesting season makes the crop not very suitable for dedicated mill and distillery production (which would be idle most of the year),

Table 11.7 Sugarcane production in South Africa

	Unit	1980	1990	2000	2010
Harvested area	million ha	0.2	0.3	0.3	0.4
Production	million tonne	14.1	18.1	23.9	16.0
Yield (three-year average)	tonne/ha	75.5	71.0	72.4	60.6
Average annual increase[1]	kg/ha/year		−447	−323	−857

[1]Figure for 1980–1990 is presented in the column for 1990; figure for 1990–2000 is presented in the column for 2000, and so on.

Sources: Calculated from FAOSTAT (2010–2013), http://faostat.fao.org; FAPRI-ISU (2011).

Table 11.8 Sugarcane use for ethanol production in South Africa

	Unit	2000	2010	2020
Use in ethanol	million tonne	Negligible	0.2	4.9
Share of all sugarcane	%	Negligible	1.0%	No data
Biofuel feedstock area	million ha	Negligible	No data	No data

Sources: Calculated from OECD-FAO (2010); Burrell (2010).

but sweet sorghum is particularly promising to complement other feedstocks such as sugarcane; for example:

• During the off season
• In 'enclaves' not suited for sugarcane production (e.g., frost pockets, drought-prone areas)
• Grown by small-scale growers in areas where sugarcane is mainly produced as a large plantation crop

Sweet sorghum is also promising for small-scale ethanol production in remote rural areas for local supply (e.g., of ethanol gel). In the southern African region, Zambia, Mozambique, and Malawi may have a high potential for sweet sorghum–based ethanol production. Agronomic and industrial trials in Europe, Asia, and Africa have demonstrated the productive potential of sweet sorghum as part of new bioenergy systems that integrate sweet sorghum with sugarcane to supply both ethanol and electricity. Bagasse (fibrous residues obtained from the stems) can be burnt to produce electricity and to process heat and power (Elbehri et al., 2013). Alternatively, it can serve as cattle feed because it is more nutritious than sugarcane bagasse.

Many aspects of sweet sorghum production are still poorly understood, such as the yield penalty quantified when grown under suboptimal conditions and what yields can be obtained practically in resource-poor environments under small-scale production systems. Research institutes develop varieties that can be grown throughout the year in order to include the crop in a variety of cropping systems. Another issue is the lack of market infrastructure for grains other than maize in Mozambique.

Jatropha for biodiesel

Jatropha (*Jatropha curcas* L.) is a perennial shrub or small tree that grows under (sub)tropical conditions and is able to withstand severe drought and low soil fertility. There are numerous documented uses of Jatropha, including in traditional medicine, lighting, soap making, and live fencing. The seeds have an oil content of approximately 35%. This oil is mainly used as a feedstock for soap production but can be used for biodiesel production as well (El Bassam, 2010; FACT, 2010).

Jatropha that originates from Central America and the northern parts of South America is presently grown in tropical areas worldwide (sub-Saharan Africa, Southeast Asia, and India) (FACT, 2010). Currently about 250 Jatropha projects are being implemented globally, covering a total area of around 900,000 ha. The bulk of this area is concentrated in Asia (85%), followed by Africa (13%) and Latin America (2%) (Elbehri et al., 2013).

Traditionally, wild-growing Jatropha in Mozambique is used as a medicinal plant, for decorative purposes, or to establish boundaries between smallholder plots. There are numerous plantation-scale projects in place in various parts of the country, including the Zambézia and Inhambane provinces (Econergy, 2008). In 2004 farmers were encouraged to produce Jatropha on unused, marginal land to make Mozambique an oil-exporting country. However, farmers who produced seeds could not sell them because markets and supply chains were absent. Plantations were established, but information on crop varieties, agronomy, markets, and scale of operations were lacking. Investments were made assuming that Jatropha would hardly require nutrients or water, but in practice land often appeared to be unsuitable (Schut et al., 2010). Detailed production data could not be obtained.

11.5 Crop cultivation

Sugarcane

Sugarcane is a semi-perennial crop which is harvested a number of times before it has to be reestablished. The first crop after planting is called the *plant crop*; subsequent crops are called *ratoons*. In South Africa farmers typically reestablish approximately 10% of their area under sugarcane annually. This is much less than in Brazil (about one-sixth) and Australia (one-third), because much of the work is done manually, which implies less soil compaction and less damage to the remaining stools after ratooning. It could also be explained because varieties in South Africa are more robust.

Activity	Jan	Feb	Mar	Apr	May	Jun	Jul	Aug	Sep	Oct	Nov	Dec
Planting	X	X										X
Growing period	X	X	X	X	X	X	X	X	X	X	X	X
Harvesting					X	X	X	X	X	X		
Post-harvest	X										X	X

Figure 11.1 Sugarcane cropping calendar for Mozambique

Irrigation is essential where the mean annual rainfall (MAR) is less than 800 mm. Although most of the commercial cane production areas in Malawi, Mozambique, Swaziland, Tanzania, and Zambia receive a MAR greater than this threshold, irrigation is routinely carried out. The lower reaches of the Zambezi in Mozambique are identified as suitable to very suitable for rain-fed sugarcane production (Watson et al., 2008).

More than 90% of cane in southern Africa is harvested manually, and more than 80% is burnt before harvesting. Harvesting and delivery of sugarcane in southern Africa account for 30% of the cost of sugar production, with transport about 15–20%. The logistic chain steps such as burning, cutting, loading, and transporting have generally been managed separately, but efficiency can be improved substantially by integration and central coordination (Bezuidenhout, 2010). Figure 11.1 presents a typical cropping calendar for Mozambique.

There are several reasons for the low yields in Mozambique. Environmental factors, especially the types of soils and the topography, are often less than optimal. Often, soils are heavy, black clay soils that 'crack' on drying and swell when wet, presenting poor drainage, which makes the soil prone to the accumulation of water, leading to problems with salinization in some areas (Econergy, 2008). The major constraints to yield increases include irrigation system efficiency, drainage, and land salinity, as well as operational issues such as the purchase of agro-chemicals and spare parts in the domestic market, power failures, and availability of labour (Econergy, 2008).Of the 432,000 ha under sugarcane in South Africa, some 18% are irrigated, mostly in the Mpumalanga province and to a lesser extent also in northern KwaZulu-Natal (Pongola and Umfolozi Flats). Sugarcane (mostly rain fed) is the major crop in KwaZulu-Natal. It is cultivated in the coastal areas along the Indian Ocean and in the Midlands region. The average sugarcane crop cycle varies between 12 and 24 months, depending on the bioclimatic region in which it is grown. In the cooler production areas of the Midlands region, at high altitude, sugarcane is harvested 18–24 months after resprouting. In the warmer coastal areas and the irrigated areas in the north, it is harvested at an average age of approximately 12 months. Cane is harvested from April to December (Department of Agriculture, Forestry and Fisheries, 2013), with most (over 85%) of the sugarcane being burnt before harvest and harvested manually.

In South Africa sugarcane accounts for 18% of fertiliser use, second after maize. About 95% of the cane crop is fertilised; typical rates are N at 92 kg per ha, P_2O_5 at 57 kg per ha, and K_2O at 133 kg per ha (FAO, 2005). In recent decades, nitrogen use efficiency has improved substantially: per tonne of cane produced, nitrogen usage decreased from an average 2.0 kg per tonne sugarcane in 1980 to about 1.5 kg without compromising yield levels (Meyer et al., 2007).

Sweet sorghum

Sorghum is grown between 40°N and 40°S and can be cultivated at elevations up to 1,500 metres. Most east African sorghum is grown between the altitudes of 900–1,500 metres. It can be grown at temperatures ranging from 15°C to 37°C with an optimum range for growth and photosynthesis of 32°C to 34°C, and day length of 10–14 hours. Optimum rainfall is between 550 and 800 mm per growing season and relative humidity is 15–50% (Rao et al., 2009).

The crop grows well on alfisols (red) or vertisols (black clay loamy) with pH 6.5–7.5, an organic matter content of more than 0.6%, and a depth of more than 80 cm. Sorghum is susceptible to sustained flooding, but will survive temporary water logging better than maize (Rao et al., 2009). While sorghum will survive up to 100 days with a supply of less than 300 mm of rainfall over the season, it responds well to additional rainfall or irrigation water. According to ICRISAT (2008) a crop of sweet sorghum with a growing period of four months will yield 2.0 tonnes of grains and 35 tonnes of green stalk cane per hectare. Grain, stalk, and residue yield, respectively, 760, 1,400, and 1,000 litres of ethanol per ha, totalling 3,160 litres of ethanol per ha.

Although sorghum is a dryland crop, sufficient moisture availability for plant growth is critically important for high yields. The great advantage of sorghum is that it can become dormant, especially in the vegetative phase, under adverse conditions and can resume growth after relatively severe drought. Early drought stops growth before panicle initiation and the plant remains vegetative; it will resume leaf production and flower when conditions again become favourable for growth. Mid-season drought stops leaf development (Rao et al., 2009). Traditional farmers use photoperiod-sensitive varieties that flower and mature during the same months regardless of planting date, so even with delayed sowing, plants mature before soil moisture is depleted (Rao et al., 2009).

In order to realize a good stalk yield, the crop should be planted at the start of the rainy season. However, this may cause competition in labour with the maize crop. Farmers who cultivate maize and traditional sorghum have adopted relay intercropping or delayed the sowing of the sorghum, resulting in lower stalk yields. Rainfed conditions in Mozambique allow only one harvest per year.

Major constraints for sweet sorghum are: poor soils coupled with improper management practices; lack of high-yielding genotypes adapted to biotic and abiotic stresses such as Striga, shoot fly, stem borer, shoot bug, aphids, anthracnose, grain mould, and leaf blight; and also lodging, drought, salinity, low temperatures, and so on (Rao et al., 2009).

The incidence of pests and diseases is bound to aggravate with increased sweet sorghum cultivation, especially if serial planting is practiced. Some of the pest-resistant material developed earlier in ICRISAT (ICSR 93034, ICSV 700 and ICSV 93046) and DSR/AICSIP (CSH22 SS, CSV19 SS and SSV74) was promising for stalk and sugar yields. It is imperative to develop resistance to key pests and diseases with improved stalk yields and sugar content (Rao et al., 2009). Plants should be processed quickly after harvesting.

Jatropha cultivation in Mozambique

Jatropha grows in (sub)tropical regions between 30°N and 35°S (FAO, 2010). Optimum rainfall for seed production varies between 1,000 and 1,500 mm (Daey Ouwens et al., 2007); at least 600 mm is needed to allow flowering and fruit setting. In areas of unimodal rainfall, flowering is continuous throughout most of the year. Optimum temperatures are between 20°C and 28°C, with no frost. The plant is well adapted to conditions of high light intensity and is unsuited to growing in shade (Britaine and Lutaladio, 2010).

The best soils for Jatropha are aerated sands and loams of at least 45 cm depth. Heavy clay soils are less suitable and should be avoided, particularly where drainage is impaired, because Jatropha is intolerant of waterlogged conditions. Ability to grow in alkaline soils has been widely reported, but the soil pH should be between 6.0 and 8.0/8.5 (Daey Ouwens et al., 2007).

Propagation is from seed or cuttings. Seeds should be sown three months before the start of the rains in polyethylene bags or tubes. The seedlings may require irrigation for the first two to three months after planting. The advantage of using cuttings is their genetic uniformity, rapid establishment, and early yield. Jatropha is suitable for intercropping, especially during the first two to five years before it starts to set fruit. For a living hedge, cuttings should be inserted between 5 and 25 cm apart and 20 cm into the ground two to three months before the onset of the rainy season (Britaine and Lutaladio, 2010).

Before the ground becomes shaded by the developing leaf canopy, it is important to control competing weeds regularly. When fully grown, it reaches a height of 3–4 metres or even up to 6 metres. Pruning during the dry season is important in order to increase branching and inflorescences and for easy harvesting. Jatropha is adapted to growing in poor soils, but a productive crop requires sufficient fertilisation and adequate rainfall or irrigation. No information on fertilisation is available, but 1 tonne of seeds respectively removes 30 kg of nitrogen, 4 kg of phosphorous, and 30 kg of potassium (Jongschaap, 2010).

Because systematic recording of yields started only relatively recently, few yield data are available. Jatropha shows a high yield variability among individual trees (Britaine and Lutaladio, 2010), with yield estimates depending on the site and the growth conditions. Achten et al. (2008) estimated a dry seed output of 1–2.5 tonnes per ha per year for degraded land and low input availability. Fertile soils and higher input levels would support yields up to 7 tonnes per ha (Daey Ouwens et al., 2007). Jongschaap et al. (2007) calculated a theoretical potential seed yield of 7.8 tonnes per ha under optimal conditions. The economic life of a Jatropha plantation reportedly ranges from 30 to 50 years (Britaine and Lutaladio, 2010).

Information will be used here from smallholder projects in Mozambique's Cabo Delgado Province with an average rainfall of 800 mm. Soils are shallow cambisols and acrisols, with waterlogging common in low-lying areas during the rainy season. Farming is mainly for self-subsistence. More than 600,000 Jatropha plants were established, with farmers preferring wide spacing in hedges around maize and cassava fields to serve as boundary markers and as a cash crop. Cuttings yield sooner than seedlings but do not develop tap roots, which makes them more susceptible to drought and prevents uptake of water and nutrients from deep soil layers (Nielsen et al., 2011).

No external inputs were used, with Jatropha production requiring only labour for pruning, weeding, and harvesting. Pruning is generally carried out during the dry season. In Cabo Delgado, pests (mostly termites) and diseases are not of economic significance. Elsewhere in Mozambique, where rainfall is higher, pest problems can be severe.

Jatropha is harvested mainly at the end of the rainy season (i.e., around March), which coincides with the harvest of food crops. Seeds can be kept on the plants for several weeks with minimal loss. A second harvest peak occurs in October just before the rains. Where water is available, smaller amounts of seeds may be produced year round (Nielsen et al., 2011). Seeds are harvested by hand, usually 1–3 kg of seeds per hour including shelling, but this can increase if yields go up (Nielsen et al., 2011).

11.6 Conversion to biofuels

The sugarcane conversion process is presented in Chapter 6. For southern Africa, fixed conversion efficiencies of 90 and 60 litres of ethanol per tonne of fresh stalks of sugarcane and sweet sorghum, respectively, are used. The energy requirement for processing is assumed to be covered by burning bagasse (i.e., the biomass remaining after stalks are crushed to extract their juice).

The oil content of Jatropha seed can range from 18% to 42%, but generally lies in the range of 30–35%, by weight. The seed kernel contains predominantly crude fat (oil) and protein, while the seed coat contains mainly fibre. For oil extraction, the nuts are milled and pressed. Small-scale, hand-operated expellers can extract 1 litre of oil for every 5.0–5.5 kg of seed. This system, which is relatively inefficient, extracts only 60% of the available oil. Engine-driven screw presses extract 75–80% of the oil, producing 1 litre of oil from every 4 kg of dried seed. Preheating, repeat pressings, and solvent extraction can enhance extraction efficiency (Britaine and Lutaladio, 2010).

To improve storage opportunities, solids must be removed from the oil. Purified oil is used as a fuel for stationary engines, requiring no processing before use but careful filtering. Heat of combustion is 39.6 MJ/kg (El Bassam, 2010).

11.7 Evaluation

Biofuel production

Biofuel production in Mozambique and South Africa so far is modest to low. Ethanol production generally is 23 and 15 million litres per year while biodiesel generated does not exceed 50 million litres. An increase has been projected, but production is expected to remain modest.

Changes in land use

Area used for biofuel production does not exceed 150,000 ha, and net biofuel area is 10,000 ha at the most (Table 11.9). Other land-use dynamics in Mozambique and South Africa show opposite trends. Mozambique has seen an expansion of agricultural area with a synchronous increase of the MCI.

Table 11.9 Changes in land availability for food and feed production in southern Africa from 2000 to 2010

	Unit	Mozambique	South Africa
Increase in biofuel production	million ha	0.13	0.12
Co-products	million ha	0.03	0.04
Net biofuel area expansion	million ha	0.10	0.08
Change in agricultural area	million ha	1.3	–2.7
Change in MCI	million ha	0.9	–0.8
Land gains total	million ha	2.1	–3.7
Net land balance	million ha	2.0	–3.7

Sources: Calculated from Tables 11.6, 11.8 and FAOSTAT (2010–2013), http://faostat.fao.org.

Net land availability for food and feed markets has increased to two million ha. In South Africa both agricultural area and harvest frequency are declining. Net availability for food production has declined. Future expansion of biofuels is not expected to change this picture dramatically.

Production efficiency

Data on sweet sorghum and Jatropha fuel production are insufficient for a detailed analysis. Sugarcane ethanol in Mozambique and South Africa is less efficient than its counterpart in Brazil, especially in terms of nutrient and water efficiency (Table 11.10). This is caused by efficient utilisation of vinasse (returning a large share of the nutrients to the field) in Brazil, as well as higher water use in southern Africa where sugarcane is commonly irrigated. Preharvest burning of leaves, which is still mainstream in southern Africa, but strongly restricted in Brazil, negatively affects the organic matter balance.

Jatropha yields on degraded land in a low-input system range from 0.5 to 2.5 tonnes of dry seeds per ha. Based on data from Jongschaap (2007), harvesting 1 tonne of seeds respectively removes 30 kg of nitrogen, 4 kg of phosphorous, and 30 kg of potassium. With recovery at 50% for nitrogen, 10% for phosphorous, and 40% for potassium, annual fertilizer gifts of 60 kg nitrogen, 41 kg phosphorous, and 75 kg potassium are required per ha. Returning presscake and prunings saves fertilizer gifts.

GHG emissions

GHG emissions occurring during crop cultivation and conversion are expected to be somewhat higher than those in Brazil, mainly due to higher (nitrogen) fertiliser application in sugarcane production and less efficient use of heat and power co-generated during processing.

Table 11.10 Performance and impacts of sugarcane ethanol production in southern Africa

	Unit	Mozambique	South Africa
Crop yield	tonne/ha	70	60
Biofuel yield	litre/ha	6,300	5,400
Biofuel yield	GJ/ha	134	114
Nitrogen productivity (crop)	kg dm/kg N	196	168
Water requirement (biofuel)	m^3/GJ	60	50
Nutrient output (biofuel)	GJ/kg N	1.3	1.1
	GJ/kg P_2O_5	3.3	2.9
Impact on soil organic matter	–	Slightly negative	Slightly negative
Risk of leaching	–	Low	Low

Economic and social impacts

Biofuel production levels still are relatively low, and no large impacts on the economy or employment opportunities are expected. Local biofuel production could, however, contribute to improved living conditions and energy security in relatively remote and isolated areas where fossil fuels usually are difficult to obtain and only at high costs.

Changes in biomass availability

Expansion of biofuel production has had no significant influence on biomass availability so far. Arable crop area expanded in Mozambique, while output per ha of arable land increased. Net biomass availability increased by three million tonnes (Table 11.11). The situation in South Africa is that crop area and output per ha are leading to a loss of biomass availability of two million tonnes. Biofuel production is playing a relatively modest role here (contributing 14% of reduced biomass availability). It is not clear what causes declining land use or yields.

The outcomes are somewhat unexpected. They suggest that the food versus fuel discussion is not relevant in these countries, which is partly explained by the low biofuel production levels that have been reported so far. Some (De Vries et al., 2012) project that although investments may help with growth enhancement and poverty reduction, biofuels may cause a rise in food prices due to competition for land and labour. This is, however, not expected in the near future.

Policy implications

Current policy targets offer no threat to food production. At the local level, however, tensions and social conflict occur when large plantations are imposed on areas of extensive forms of land use and/or if water abstraction affect the livelihoods of local populations. Even if restricted and localized, such incidents may contribute to widespread perceptions of insecurity

Table 11.11 Changes in food biomass availability in southern Africa 2000–2010

	Unit	Mozambique	South Africa
Increase total biomass production	million tonne	4.0	−3.8
Feedstocks used for biodiesel production	million tonne	0.4	0.4
Generation of co-products	million tonne	0.1	0.1
Net impact of biofuels	million tonne	0.3	0.3
Net change in biomass availability	million tonne	4.0	−4.1
Increase in biomass productivity (BP)	million tonne	6.3	−0.04

Source: Calculated from FAOSTAT (2010–2013), http://faostat.fao.org.

regarding land tenure and access to water resources. Mozambique seems to be able to profit most from small-scale biofuel production, which poses fewer threats to the position of smallholders and the local population.

11.8 Conclusion

The production of biofuels in southern Africa, currently still in its infancy, offers considerable potential benefits to the region, including improved energy security, employment, and income generation. Thus far, however, little of these promises are realized. Even though current policy targets offer no threat to food production, tensions at the local level may result when large plantations are imposed on existing extensive forms of land use. Mozambique seems to be in relatively good position to host additional bio-fuel production units, provided that the position of the existing population is guarded. Under conditions of water constraints, sweet sorghum and – to a lesser extent – Jatropha can offer good for sugarcane. Current sweet sor-ghum varieties, however, do not provide sufficient long harvesting options to allow successful operation of ethanol plants.

References

Achten, W.M.J., Verchot, L., Franken, Y.J., Mathijs, E., Singh, V.P., Aerts, R., and Muys, B. (2008). Jatropha bio-diesel production and use. *Biomass and Bioenergy*, Vol 32, pp1063–1084.

Alfani, F., Azzarri, C., d'Errico, M., and Molini, V. (2012). *Poverty in Mozambique: New evidence from recent household surveys.* The World Bank. Retrieved 19 February 2013 from https://openknowledge.worldbank.org/bitstream/handle/10986/12050/wps6217.pdf?sequence=1

Bernd, F., Reinhardt, G., Malavelle, J., Faaij, A., and Fritsche, U. (2012). *Global assessments and guidelines for sustainable liquid biofuels.* A GEF Targeted Research Project. Heidelberg, Paris, Utrecht, Darmstadt.

Bezuidenhout, C.N. (2010). Review of sugarcane material handling from an inte-grated supply chain perspective. *Proceedings of the South African Sugar Technolo-gists' Association*, Vol 83, pp63–66. Retrieved from www.sasta.co.za/wp-content/uploads/Proceedings/2010s/2010-Bezuidenhout.pdf

Britaine, R., and Lutaladio, N. (2010). Jatropha: A smallholder bioenergy crop – The potential for pro-poor development. In *Integrated Crop Management*, Vol 8. Rome, Italy: Food and Agricultural Organization of the United Nations.

Burrell, A. (Ed.). (2010). *Impact of EU biofuel target on agricultural markets and land use.* Retrieved 12 January 2012 from http://publications.jrc.ec.europa.eu/repository/bitstream/111111111/15287/1/jrc58484.pdf

Cartwright, A. (2010). Biofuels trade and sustainable development: The case of South African bioethanol. In A. Dufey and M. Grieg-Gran (Eds.), *Biofuels pro-duction trade and sustainable development* (pp. 65–102). London: IIED.

Cheeseman, O. (2004). *Environmental impacts of sugar production.* Wallingford, UK: CABI Publishing.

Daey Ouwens, K., Francis, G., Franken, Y.J., Rijssenbeek, W., Riedacker, A., Foild, N., Jongschaap, R.E.E., and Bindraban, P. (2007). State of the art, small and large scale project: Position paper on Jatropha curcas. In *Position paper on Jatropha and large scale project development*. Wageningen, the Netherlands: FACT Foundation. Paper presented at the Expert seminar on Jatropha curcas L. Agronomy and genetics, Wageningen, the Netherlands, 2007-03-26/ 2007-03-28

Department of Agriculture, Forestry and Fisheries. (2013). *Trends in the agricultural sector 2012*. Pretoria, South Africa: Department of Agriculture, Forestry and Fisheries.

De Vries, S. C., Van de Ven, G.W.J., Van Ittersum, M. K., and Giller, K. E. (2012). The production-ecological sustainability of cassava, sugarcane and sweet sorghum cultivation for bioethanol in Mozambique. *Global Change Bioenergy*, Vol 4, pp20–35.

DME. (2003). *White paper on renewable energy*. Retrieved from http://unfccc.int/files/meetings/seminar/application/pdf/sem_sup1_south_africa.pdf

DME. (2007). *Biofuels industrial strategy of the Republic of South Africa*. Retrieved from http://www.info.gov.za/view/DownloadFileAction?id = 77830

Econergy. (2008). *Mozambique biofuels assessment*. Report prepared for Mozambique's Ministries of Agriculture and Energy and the World Bank, Washington, DC.

EIA. (2013). USA Energy Information Administration. Retrieved 1 February 2013 from http://www.eia.gov/cfapps/ipdbproject/iedindex3.cfm?tid=79&pid= 80&aid=1&cid = regions&syid = 2000&eyid = 2011&unit = TBPD

El Bassam, N. (2010). *Handbook of bioenergy crops: A complete reference to species, development and applications*. London: Earthscan.

Elbehri, A., Segerstedt, A., and Liu, P. (2013). *Biofuels and the sustainability challenge: A global assessment of sustainability issues, trends and policies for biofuels and related feedstocks*. Rome, Italy: Food and Agricultural Organization of the United Nations.

FACT. (2010). *The Jatropha handbook, from cultivation to application*. Wageningen, the Netherlands: FACT Foundation.

FAO. (2005). *Fertilizer use by crop in South Africa*. Retrieved from ftp://ftp.fao.org/docrep/fao/008/y5998e/y5998e00.pdf

FAO. (2010). *FAO/WFP crop and food security assessment mission to Mozambique*. Rome, Italy: Food and Agricultural Organizaiton of the United Nations, World Food Programme.

FAPRI-ISU. (2011). *World agriculture outlook*. Retrieved 18 May 2012 from http://www.fapri.iastate.edu/outlook/2011/

Government of Mozambique. (2009). *Política e estratégia de biocombustíveis*, p9.

Government of Mozambique. (2011). *Biofuels blending regulation*, p6.

Government of South Africa. (2013). Retrieved 9 September 2013 from http://www.info.gov.za/aboutsa/agriculture.htm

ICRISAT. (2008). *Eastern and southern Africa region: 2007 highlights*. Nairobi, Kenya: ICRISAT (International Crops Research Institute for the Semi-Arid Tropics).

INE. (2011). *Censo agro: Pecuário 2009–2010: Resultados definitivos*. Moçambique: Instituto Nacional de Estatística.

INE. (2013). Instituto Nacional de Estatística. Retrieved 4 April 2013 from http://www.ine.gov.mz/

Janssen, R., and Rutz, D. (2012). Keynote introduction: Overview on bioenergy policies in Africa. In R. Janssen and D. Rutz (Eds.), *Bioenergy for sustainable development in Africa* (pp. 165–182). Dordrecht, the Netherlands: Springer. DOI: 10.1007/978–94–007–2181–4_14

Jones, S., and Tarp, F. (2012). *Jobs and welfare in Mozambique.* Washington, DC: The World Bank.

Jongschaap, R.E.E. (2010). Chapter 4. Agronomy aspects. In Eijk, J. van, Smeets, E., Jongschaap, R.E.E., Romijn, H., Balkema, A. (Eds.) *Jatropha Assessment. Agronomy, socio-economic issues, and ecology. Facts from literature.* Utrecht, the Netherlands: AgentschapNL. pp. 22–35.

Jongschaap, R.E.E., Corré, W.J., Bindraban, P.S., and Brandenburg, W.A. (2007). *Claims and facts on* Jatropha curcas L. Wageningen, the Netherlands: Plant Research International.

Leete, M., Damen, B., and Rossi, A. (2013). *Mozambique BEFS country brief.* Rome, Italy: FAO.

Letete, T., and Von Blottnitz, H. (2012). Biofuel policy in South Africa. In R. Janssen and D. Rutz (Eds.), *Bioenergy for sustainable development in Africa* (pp. 191–199). Dordrecht, the Netherlands: Springer. DOI: 10.1007/978–94–007–2181–4_14

Meyer, J.H., Schumann, A.W., Wood, R.A., Nixon, D., and van den Berg, M. (2007). Recent advances to improve nitrogen use efficiency of sugarcane in the South African sugar industry. *Proceeding of the International Society of Sugar Cane Technologists,* Vol 26, pp238–246.

Nielsen, F., De Jongh, B., and de Jongh, J. (2011). *End report FACT pilot project: Jatropha oil for local development in Mozambique 2007–2010.* Wageningen, the Netherlands: FACT Foundation.

NL Agency. (2012). *Mozambique: Market opportunities for bioenergy.* Utrecht, the Netherlands: NL Agency.

OECD- FAO. (2009). *Agricultural outlook 2009–2018.* Paris, France: Organisation for Economic Co- operation and Development; Rome, Italy: Food and Agricultural Organization of the United Nations.

OECD-FAO. (2010). *Agricultural outlook 2010–2019.* Paris, France: Organisation for Economic Co-operation and Development; Rome, Italy: Food and Agricultural Organization of the United Nations.

Rao, S.P., Rao, S.S., Seetharama, N., Umakath, A.V., Reddy, P.S., Reddy, B.V.S., and Gowda, C.L.L. (2009). Sweet sorghum for biofuel and strategies for its improvement. *International Crops Research Institute for the Semi-Arid Tropics, Information Bulletin,* Vol 77. Retrieved from http://oar.icrisat.org/1354/

Schut, M., Slingerland, M., and Locke, A. (2010). Biofuel developments in Mozambique: Update and analysis of policy, potential and reality. *Energy Policy,* Vol 38, pp5151–5165. DOI: /10.1016/j.enpol.2010.04.048

Watson, H.K., Garland, G.G., Purchase B., Dercas, N., Griffee, P., and Johnson, F.X. (2008). *Bioenergy for sustainable development and global competiveness: The case of sugar cane in southern Africa.* A compilation of results from the Thematic Research Network: Cane Resources Network for Southern Africa (CARENSA). Stockholm, Sweden: Stockholm Environmental Institute.

World Bank. (2013a). *Agriculture, value added (% of GDP).* Retrieved 9 September 2013 from http://data.worldbank.org/indicator/NV.AGR.TOTL.ZS/countries

World Bank. (2013b). *South Africa overview.* Retrieved 23 March 2013 from http://www.worldbank.org/en/country/southafrica/overview

12 Biofuel production in China

*J.W.A. Langeveld, J. Dixon and
H. van Keulen*

12.1 Introduction

Since the introduction of the household responsibility system reforms in 1978, China has been transforming from a centrally planned towards a market economy and has been experiencing rapid economic and social development. An annual gross domestic product (GDP) growth of 10% during the past decade has lifted more than 600 million people out of poverty. With a population of 1.3 billion, China now is the second-largest economy in the world. In other respects, however, it still remains a developing country, and its market reforms have not yet been completed (World Bank, 2013).

In 2011 Chinese per capita income ranked 114th in the world, and more than 170 million people still live below the $1.25-a-day poverty line. With the second-largest number of poor in the world, poverty reduction remains a challenge while rapid economic growth has introduced challenges such as high income inequality, rapid urbanization, and declining environmental sustainability. China also faces demographic pressures related to an ageing population and internal labour migration. Policy adjustments, needed to sustain economic growth, are defined in China's most recent (12th) Five-Year Plan (2011–2015), which addresses social imbalances, access to education, pollution, and energy efficiency (World Bank, 2013).

Meanwhile, economic growth has led to a tremendous increase in energy demand. Using fossil fuels has had serious implications for greenhouse gas (GHG) emissions. Compared to 1990, the amount of CO_2 released in China in 2004 had doubled while the corresponding amount per person increased by 86%. This is in strong contrast with global trends: world GHG emissions increased by some 28%, and per-person emissions increased by only 5% during the same period (Zhong et al., 2010).

Grain self-sufficiency remains one of the most important agricultural policy goals. China is feeding 22% of the world's population on only 7% of its land area, achieved with only modest imports. However, a high price has been paid for this achievement, where yield increases rely heavily on intensive use of external inputs. Intensive farming systems have contributed to soil degradation, water scarcity, and pollution, as well as to declining efficiency of fertilizer application (Youa et al., 2011).

   The central role of grain security and agricultural production was confirmed in the first policy document for 2013, stating that China should improve the development of agricultural production. The document, issued by the central committee of the Communist Party and the State Council, is the tenth consecutive annual text focusing on rural issues (BioenergySite, 2013).

   China will continue to increase grain output by maintaining the area sown to grain crops, improving production structure, and raising per-unit yields. Efforts will be made to strengthen agricultural infrastructure and increase production efficiency, while strict rules on farmland protection will be maintained. Efforts should be expanded to improve irrigation facilities and water conservation and to encourage innovation in agriculture.

   To encourage farm production, China aims to increase the minimum purchase price for wheat and rice and launch a program for the purchase and storage of corn, soybeans, rapeseed, cotton, and sugar. Improvements are required in agricultural market supervision and early warning systems to better regulate the domestic market, as well as the import and export of agricultural products (BioenergySite, 2013).

   This chapter describes biofuel policies and production in the light of a productive food system that depends on the high use of external inputs. The main focus is on biofuel production and its relation to land use and agricultural output. After describing land resources and land use in China (Section 12.2), this chapter discusses biofuel policy and industrial development (Section 12.3). Section 12.4 defines feedstock requirements for ethanol production, while Sections 12.5 and 12.6 elaborate on the production of major ethanol feedstock crops (corn and wheat) and their conversion to ethanol. This is followed by an assessment (Section 12.7) and some conclusions (Section 12.8).

12.2 Land resources

China's total land area is 930 million ha. Of this, more than half is agricultural area. Forest and 'other' land each cover some 20% (Figure 12.1).

   The climate of China varies greatly, from tropical in the south to subarctic in the north. It is dominated by monsoon winds that are caused by differences in the heat-absorbing capacity of the land and ocean. Alternating seasonal air-mass movements and accompanying winds are moist in summer and dry in winter. Monsoons govern the timing of the rainy season and amounts of rainfall. Tremendous differences in latitude, longitude, and altitude result in strong variations in precipitation and temperature. Although most of the agricultural land lies in the temperate belt, its climatic patterns are complex ('Geography of China', 2013).

   Agricultural area amounts to 524 million ha (Table 12.1), mainly permanent grassland (76%). Some 20% is cultivated with arable crops; tree crops occupy 3% (14 million ha). Data for fallow or fodder crops are scarce, but

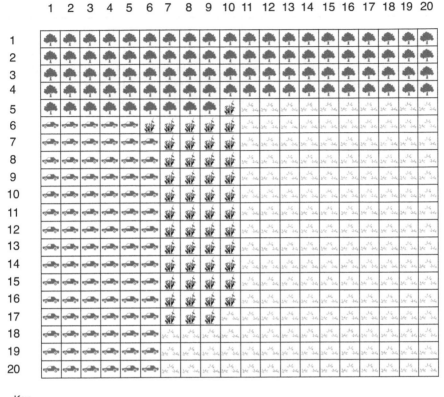

Key

forest   permanent grassland   arable crops

other

Figure 12.1 Schematic overview of land use in China

No geographical representation. Each cell represents 0.25% of land area. Position of the categories was chosen randomly.

Source: Calculated from FAOSTAT (2010–2013), http://faostat.fao.org.

fodder crop area was estimated at 2% of the total land area (Simpson et al., 1994). The Multiple Cropping Index (MCI) is 1.5, indicating that each ha of agricultural land on average is harvested 1.5 times per year. This high intensity is related to cultivation of tropical crops (especially rice), which have a growth cycle of less than four months, in combination with favourable climatic conditions and intensive land-use practices.

Following the introduction of national reforestation policies, forest area in China has increased by 30% since 1990. This is mostly at the expense of 'other' land, most likely land that formerly was abandoned plus (non-forest) nature area. Agricultural area increased in the 1980s, but is presently

Table 12.1 Land area and land use in China

	Unit	1980	1990	2000	2010
Forest area	million ha	No data	157	177	207
Other land	million ha	No data	244	224	207
Agricultural area	million ha	434	531	532	519
Permanent grassland	million ha	334	400	400	393
Agricultural tree crops	million ha	3	8	11	15
Arable area	million ha	97	124	121	111
Of which:					
arable crop area	million ha	83	105	103	95
fallow	million ha	14.5	18.6	18.1	16.7
Multiple Cropping Index (MCI)	–		1.39	1.22	1.33	1.53

Source: Calculated from FAOSTAT (2010–2013), http://faostat.fao.org.

declining, probably giving way to urbanization and industrialization. So far, the loss of arable land has been, however, more than compensated by intensification of land use on the remaining agricultural area. Since 1990, MCI increased by 25%. The grassland area seems to be constant, whereas the agricultural tree crop area is growing.

12.3 Markets and policy conditions

China has a long history of liquor brewing and utilization of oil and fat. Until the end of the twentieth century, fuel ethanol and biodiesel primarily were produced in China from aged grains and oils. Driven largely by the soaring oil demand and the consequently increased imports, China has been promoting biofuel production since 2002. Strong governmental support, including direct subsidies, tax exemption, and low interest loans, helped to develop China's biofuel industry (Zhong et al., 2010).

Ethanol production capacity reached 2.4 million litres by 2008, with four of the five licensed fuel ethanol plants in operation using corn and wheat as feedstocks and the remaining one using cassava. Total ethanol fuel production of China was 1.9 million litres in 2007, accounting for 20% of national gasoline consumption that year (Zhong et al., 2010). Targets set for fuel ethanol and biodiesel production in the Medium- and Long-Term Development Plan for Renewable Energy in China are 12.6 and 2.5 million litres by 2020, respectively. These official targets were considered to be conservative and may very well be exceeded (Yan et al., 2010).

Following early biofuel production developments, the Chinese government announced its intention to halt building grain-based ethanol plants and to develop methods that use non-food crops as feedstocks. This is based on the potential threat that fuel conversion poses on the national food security (Yan et al., 2010). The Ministry of Agriculture of the People's Republic of

China's agricultural biomass energy industry development plan (2007–2015) explicitly states that fuel ethanol development in China should not compete with food crops or limit land used for food production. Crops considered as 'non-food' include feed and industrial crops such as cassava and sweet sorghum, while cellulosic biomass and algae are considered as new resources that can provide raw materials for China's fuel development (Zhong et al., 2010).

China's State Council recently published a white paper specifying that non-fossil fuels should make up 11% of the primary energy in 2015 (end of the 12th Five-Year Plan). Renewables should generate 30% of the energy production, but production should be adjusted to local conditions using a variety of feedstocks, including crop residues, grain-processing residues, and bagasse. Woody biomass energy will be produced in forested areas, while electricity will be generated from waste incineration and landfill gas. The construction of biogas infrastructure will be accelerated, and development of production capacity for biodiesel and cellulosic ethanol will be supported (Voegele, 2013).

International Energy Agency (IEA) data suggest that in 2009 solid biofuels accounted for 2,350 GWh of gross electricity generation. During the same period, biomass accounted for 11,900 GWh in gross heat production. Recent (2012) data show that non-hydro renewables made up only 0.3% of China's energy consumption in 2009. Hydroelectric and coal accounted for 6% and 70%, respectively (Voegele, 2013).

Fundamental research on biodiesel production from vegetable oil was carried out in the 1980s. In 2003 biodiesel development was listed in the National Science and Technology Industrialization Plan. In 2004 industrial exploitation of biodiesel production was supported by the Ministry of Science and Technology.

Biofuel production in China started slowly in the early 2000s. Annual ethanol production and consumption in 2010 exceeded 2 billion litres, up from 1.2 billion in 2005 (Table 12.2). Biodiesel production is only a fraction of this. For 2020 the Food and Agricultural Policy Research Institute

Table 12.2 Biofuel production and market development in China

	Unit[1]	2005	2010	2020
Ethanol production	billion litre/year	1.2	2.1	3.0
Ethanol consumption	billion litre/year	1.2	2.1	
Biodiesel production	billion litre/year	0.05	0.35	
Biodiesel consumption	billion litre/year	0.05	0.35	
Support investments	billion $	No data	No data	
Subsidies, tax exemptions	$/litre	0.40[2]	0.17	

[1]billion = thousand million
[2]Figure for 2006.

Sources: Calculated from FAPRI-ISU (2011); EIA (2013).

(FAPRI-ISU, 2011) projects a small increase in ethanol production (reaching 3 billion litres). Not much information on biofuel subsidies could be obtained. Subsidies for ethanol were US$0.40 in 2006 but were reduced to US$0.17 in 2010.

By 2020 the area of energy forests is expected to exceed 120 million ha, providing more than 2.3 million tonnes of biodiesel to meet the fuel requirements of power plants with more than 11 million kW of installed capacity. During the period of the 11th Five-Year Plan, the State Forestry Administration focused on developing 7.7 million ha of *Jatropha curcas* and 6 million ha of energy shrubs such as *Salix psammophila*, eabuckthorn, *Amorpha fruticosa*, *Robinia pseudoacacia*, Chinese tamarisk, and Quercus in North Shaanxi, East China, Central China, and collective forest areas in South China (Wang and Xiao, 2009).

Sweet sorghum is considered to be a high-potential feedstock in China's ethanol industry. A 5,000-tonne demonstration project has been established, although under current conditions the GHG footprint of sorghum ethanol may exceed that of gasoline. Cassava has been identified as one of the most promising biofuel feedstocks in China because it can be grown on mountainous land or agriculturally marginal land unsuitable for food crop production (Yan et al., 2010).

Rapeseed biodiesel has been associated with low GHG emissions, but being a large net importer of oil seeds (soybean, rapeseed), China may not be in a position to use these feedstocks for biodiesel production. Jatropha is regarded as a promising feedstock, especially because of its minimal water and nutrient requirements and its ability to grow on poor quality land.

Life-cycle studies suggest favourable GHG balances for Jatropha, which is grown in Sichuan, Yunnan, and Guizhou provinces while suitable areas for expansion are found elsewhere (e.g., Guangxi, Guangdong, and Hainan; Wang and Xiao, 2009). There are a few small-scale biodiesel plants that run on used cooking oil. The potential for this feedstock, however, is believed to be less than 2.3 million litres per year (Yan et al., 2010).

Many households that keep pigs, dairy cows, beef cattle, broilers, and sheep have access to valuable manure, while more than 300 million tonnes of stover are available for utilization. Some 146 million households have access to suitable biogas feedstocks, but only 120 million encounter favourable conditions for production. An estimated 84 million offer realistic opportunities for local energy sources. Biogas production units have been built for 18 million households (at the end of 2005; Wang and Xiao, 2009).

12.4 Feedstock requirements

Two-thirds of Chinese ethanol is produced from corn. The remainder originates from wheat; sugarcane plays no important role. The first factory that runs on cassava, sugarcane, and sugar beet was taken into production recently in Guangxi province. Sugar crops are preferred feedstocks because they require no hydrolysis and thus are cheaper to use in ethanol production.

Table 12.3 Corn production in China

	Unit	1980	1990	2000	2010
Harvested area	million ha	20	21	23	33
Production	million tonne	63	97	106	178
Yield (three-year average)	tonne/ha	3.0	4.3	4.7	5.5
Average annual increase[1]	kg/ha/year		129	42	74

[1]Figure for 1980–1990 presented in the 1990 column; figure for 1990–2000 in the 2000 column, and so on.

Source: Calculated from FAOSTAT (2010–2013), http://faostat.fao.org.

Table 12.4 Wheat production in China

	Unit	1980	1990	2000	2010
Harvested area	million ha	29	31	27	24
Production	million tonne	55	98	100	115
Yield (three-year average)	tonne/ha	2.0	3.1	3.8	4.8
Average annual increase[1]	kg/ha/year		107	72	94

[1]Figure for 1980–1990 presented in the column for 1990; figure for 1990–2000 presented in the column for 2000, and so on.

Source: Calculated from FAOSTAT (2010–2013), http://faostat.fao.org.

While great effort has been made to develop bioethanol production from sweet sorghum, production is currently still in the pilot phase (Wang et al., 2009). Information on biodiesel could not be obtained, but it is assumed that most is from soybean.

The Chinese corn area has increased by more than half since 1980, amounting to 33 million ha in 2010 (Table 12.3). Production almost tripled to reach 178 million tonnes, mostly because yield nearly doubled. Yield increases show similar trends to those of wheat, with highest values in the 1980s and 2000s, following a relatively 'slow' decade.

Wheat area declined from 29 to 24 million ha in 30 years, but production doubled, as yields increased from 2.0 to 4.8 tonnes per ha (Table 12.4). Average annual yield increase amounted to an impressive 107 kg per ha in the 1980s. It declined in the 1990s but climbed to 94 kg per ha per year during the past decade.

The quantity of grain used in ethanol production is very limited. In 2005 some 3 million tonnes of corn and wheat was used (Table 12.5), some of which was old grain storages. This use increased to nearly 6 million tonnes in 2010. As the use of food grains for ethanol is restricted by state regulations, less than 2% of wheat and 2.5% of corn is utilized. Area cultivated was 1.1 million ha in 2010. Another increase of 2 million ha is expected in 2020.

Distribution of some main feedstock conversion units in China is depicted in Figure 12.2.

Table 12.5 Wheat and corn use for ethanol production in China

	Unit	2005	2010	2020
Wheat use in ethanol	million tonne	0.9	1.7	2.3
Share of all wheat	%	0.9%	1.5%	
Corn use in ethanol	million tonne	2.2	4.2	5.5
Share of all corn	%	1.1%	2.4%	
Soybean use in biodiesel	million tonne	0.3	1.8	
Share of all soybean	%	1.5%	11.6%	
Biodiesel feedstock area[1]	million ha	0.6	1.1	

[1]Assuming yield improvement will be similar to that in the period 2000–2010.

Sources: Calculated from EIA (2013); FAOSTAT (2010–2013), http://faostat.fao.org; FAPRI-ISU (2011).

■ Corn, ◣ Rootcrop, ◆ Sugarcane, ★ Sugar beet, ● Rapeseed

Figure 12.2 Main feedstocks used in biofuel production in China
Source: Wang et al. (2009).

12.5 Crop cultivation

In contrast to previous chapters, few details are presented here as to soils in use, crop rotations, and inputs applied. Chinese cereal production has been characterised as very intensive, often receiving large amounts of fertilizers, agro-chemicals and – where applicable – irrigation water. Following Wang et al. (2012), Liu et al. (2011), Huang et al. (2012), and He et al. (2010), we estimate fertilizer applications for corn at 260 kg of N, 85 kg of P_2O_5, and 40 kg of K_2O. Wheat application levels are slightly higher for nitrogen (275 kg), lower for phosphate (50 kg), and similar for potassium oxide.

12.6 Conversion to biofuels

The principles of corn and wheat conversion have been presented in previous chapters. Conversion efficiency of wheat and corn to ethanol production in China are assumed to be similar to those reported for the USA (corn) and the EU (wheat).

12.7 Evaluation

Biofuel production

Biofuel production in China is and has been modest compared with Brazil and the USA. A production of 2 billion litres in 2010, remarkably low for such a large country, is explained by China's strong focus on primary cereal production and food availability, which allow very little room for biofuel production based on current major biofuel crops. While no major changes are expected in the near future, in the long run, availability of lignocellulosic feedstock conversion technology and production of biogas could become major sources of bioenergy.

Changes in land use

So far, the small volume of emerging biofuel production hardly has affected land availability. Since 2000, an additional 2.2 million ha has been used. Co-product generation is allocated 0.4 million ha, leaving a net area of 1.7 million ha for biofuel production. During this period, 11 million ha of agricultural land has been lost. Meanwhile, intensification (increased MCI) generated 20.3 million ha of harvested arable crop area (Table 12.6). Consequently, harvested area increased by 5.1 million ha.

According to Qu et al. (2011), decline in arable area after 2000 accelerated due to ecological recovery programs aimed to increase forest area and fight erosion. Urbanization and industrialization seemed to have played a lesser role. Still, grassland degradation and wind erosion have become much more severe. In northern China, particularly in the 3-H (Hai and Luan, Huai, and Huang) river basins, water availability has tightened.

Table 12.6 Changes in land availability in China 2000–2010

	Unit	Impact on Land Availability for Food Production	%
Increase in biofuel production	million ha	−2.2	
Co-products	million ha	0.4	
Net biofuel area	million ha	−1.8	12%
Loss of agricultural area	million ha	−13.4	88%
Total loss of available land	million ha	−15.2	100%
Change in MCI	million ha	20.3	
Net land balance	million ha	5.1	

Sources: Calculated from Tables 12.1, 12.3–12.5 and FAOSTAT (2010–2013), http://faostat .fao.org.

Groundwater tables have fallen in the Hai river basin as farmers increasingly rely on groundwater for irrigation.

Evidence on other parts of northern China is mixed. Surface water pollution locally has worsened since the beginning of the 1990s in southern China and until recently northern China. Water quality problems, to a large extent caused by agriculture-based non-point source pollution, in the larger rivers in southern China are less severe and are declining (Qu et al., 2011).

Production efficiency and GHG emissions

As no data on input use were obtained, we have no clear view of efficiency of biofuel production in China nor of its impact on soil and water resources. GHG balances could not be calculated.

Economic and social impacts

Production of biofuel feedstocks and their conversion play only a very small role in the rural economy, while the strong recent industrial development offers many opportunities for off-farm employment. National policy makers carefully monitor development of the production capacity, and it seems that priority of the biofuel sector is relatively low. More prospects can be expected when lignocellulosic conversion technologies become commercially competitive.

Development and implementation of small-scale, decentralized, biogas production units offers welcome additional sources of energy while digestate has good perspectives for application as organic fertilizers.

Biomass availability

Major efforts are being made in China to ensure increased biomass food production and agricultural output. As a result of this, net impact of the

Table 12.7 Changes in food biomass availability in China 2000–2010

	Unit	Impact on Biomass Availability
Increase total biomass production	million tonne	342
Biomass in biofuel production	million tonne	−7.3
Generation of co-products	million tonne	4.0
Net impact of biofuels	million tonne	−3.4
Net change in biomass availability	million tonne	339
Increase in biomass productivity (BP)	million tonne	412

Source: Calculated from Tables 12.3 and 12.4, and FAOSTAT (2010–2013), http://faostat.fao.org.

small biofuel industry on biomass availability is negligible. Previously we discussed that net harvested area increased by more than 5 million ha. Total biomass production in 2010 was nearly 500 million tonnes (30%) higher than arable output at the start of the twenty-first century (fodder crops not included), realizing an increase in net biomass availability of 480 million tonnes (Table 12.7).

There is little information on macro-level biomass output development, and it is not easy to put this development in perspective. The reported increase of biomass production is remarkable. An increase of 30% in a decade is no small achievement, but it is in line with earlier improvements (data not shown here). Most of this must be attributed to intensification, a combination of increasing cropping intensity, higher crop yields, and replacement of less productive crops. The output of per ha of arable land increased by 40%.

According to Wang et al. (2009), bioethanol fermentation technology from sugar feedstocks has already matured in China, but fuel bioethanol production from sugarcane and sugar beet is not feasible. This is because the demand for sugar is so great in China due to its large population, and sugarcane and sugar beet remain the main resources for sugar production. Furthermore, sugar production from sugarcane and sugar beet is currently more profitable than that of bioethanol production for Chinese corporations (Wang et al., 2009).

Policy implications

Policies with respect to biofuel production are clear and are not expected to undergo important changes over the next few years. Data presented in this chapter suggest that policies limiting the competition with food production (food cereals on the one hand and oil crops on the other hand) have been effective. They also seem to confirm the potential for use of crop residues and other lignocellulosic feedstocks, be it that no detailed quantitative analysis was done on their potential economic perspective or impacts. The relation between stover removal and soil quality needs to be carefully monitored. Potential social and economic implications of biogas production seem to require more attention, while the use of digestate as a source of nutrients and organic matter for soil improvement needs to be determined.

12.8 Conclusion

China has had modest levels of biofuel production, partly as a consequence of regulations restricting the use of food grains for biofuels. Food grain productivity has risen very fast in China, but the rate of growth is likely to soften in future decades. Input use in crop production is expected to remain very high, posing risks for soil and water resources.

This does not mean that there are no perspectives for biofuels. Development of lignocellulosic conversion technology will allow Chinese farmers to make more efficient use of crop residues while industrial waste streams and forest residues offer considerable scope for development. In addition, expansion of the biogas program can provide a cheap, robust, and safe energy source. This is expected to provide opportunities especially in remote and isolated areas, including large but extensive agricultural regions in west China.

The main research challenges for bioethanol production include (1) the use of biotechnology to introduce new characteristics to bioethanol feedstock crops – disease and pest resistance, drought and salinity tolerance, and high yield; (2) the improvement and innovation of technologies for starch-based bioethanol – such as fermentation of untreated feedstock – filtrate and recycling of clear broth, and waste treatment for bioethanol production from cassava; and (3) the development of new processes for bioethanol production from non-food-based feedstocks, such as lignocellulosic biomass (Wang et al., 2009).

For biodiesel production, the main research challenges include (1) the development and promotion of cultivars and hybrids of conventional and potential oil-producing species; (2) the development of non-edible, oil-producing plant species with high yield and good environmental adaptation; (3) the development of cultivation practices on non-edible, woody, oil-producing plants on arable and non-arable lands; and (4) the evaluation and innovation in crop production systems including harvesting of non-edible oil species (Wang et al., 2009).

References

BioenergySite. (2013). China says grain security top priority. Retrieved 5 February 2013 from http://www.thebioenergysite.com/news/12286/china-says-grain -security-top-priority

EIA. (2013). *International energy statistics*. USA Energy Information Administration. Retrieved 1 February 2013 from http://www.eia.gov/cfapps/ipdbproject/iedindex3 .cfm?tid=79&pid=80&aid=1&cid=regions&syid=2000&eyid=2011&unit= TBPD

FAPRI-ISU. (2011). *World agriculture outlook*. Retrieved 18 May 2012 from http://www.fapri.iastate.edu/outlook/2011/

Geography of China: Climate. (2013). *Wikipedia*. Retrieved 17 February 2013 from http://en.wikipedia.org/wiki/Climate_of_China#Climate

He, P., Li, S., and Jin, J. (2010). Nutrient uptake and utilization of nutrient management for maize and wheat rotation in North Central China. In *Proceedings of the International Plant Nutrition Colloquium XVI*. University of California–Davis, Davis, CA. Retrieved 22 September 2013 from http://escholarship.org/uc/ipnc_xvi

Huang, J., Xiang, C., Jia, X., and Hu, R. (2012). Impacts of training on farmers' nitrogen use in maize production in Shandong, China. *Journal of Soil and Water Conservation,* Vol 67, pp321–327.

Liu, X., He, P., and Jin, J. (2011). A long-term analysis of factors to improve nutrient management for winter wheat production in China. *Better Crops,* Vol 95, pp16–18.

Qu, F., Kuyvenhoven, A., Shia, X., and Heerink N. (2011). "Sustainable natural resource use in rural China: Recent trends and policies." *China Economic Review,* Vol 22, pp444–460.

Simpson, J.R., Cheng, X., and Miyasaki, A. (1994). *China's livestock and related agriculture: projections to 2025*. Wallingford, UK: CAB International.

Voegele, E. (2013). China offers subsidies for additional biomass power projects. *Biomass Magazine*. Retrieved 6 February 2013 from http://www.biomassmagazine.com/articles/8498/china-offers-subsidies-for-additional-biomass-power-projects

Wang, F, Xiong, X.-R., and Liu, Ch.-Z. (2009). Biofuels in China: Opportunities and challenges. *In Vitro Cellular and Developmental Biology: Plant,* Vol 45, pp342–349. DOI 10.1007/s11627–009–9209–7

Wang, M., and Xiao, M. (2009). *Second periodic activity report: Biomass resource assessment for China*. Beijing, China: Chinese Association of Rural Energy Industry (CAREI).

Wang, W., Lu, J., Ren, T., Li, Y., Zou, J., Su, W., and Li, X. (2012). Inorganic fertilizer application ensures high crop yields in modern agriculture: A large-scale field study in Central China. *Journal of Food, Agriculture and Environment,* Vol 10, pp703–709.

World Bank. (2013). Retrieved 17 February 2013 from http://www.worldbank.org/en/country/china/overview

Yan, X., Inderwildi, O. R., and King, D. A. (2010). Biofuels and synthetic fuels in the US and China: A review of well-to-wheel energy use and greenhouse gas emissions with the impact of land-use change. *Energy and Environmental Science,* Vol 3, pp190–197. DOI: 10.1039/b915801d.

Youa, L., Spoor, M., Ulimwengu, J., and Zhang, S. (2011). Land use change and environmental stress of wheat, rice and corn production in China. *China Economic Review,* Vol 22, pp461–473.

Zhong, Ch., Cao, Y.-X., Li, B.-Z., and Yuan, Y.-J. (2010). Biofuels in China: Past, present and future. *Biofuels, Bioproducts and Biorefining,* Vol 4, pp326–342. DOI: 10.1002/bbb.207.

13 Lignocellulosic crops

E. Maletta and M.V. Lasorella

13.1 Introduction

Since the advent of agriculture, more than one-fourth of Earth's land surface has been converted for agricultural purposes. Conversion from natural to agricultural landscapes dramatically changes the plant communities that are integral to ecosystem processes. Renewable biomass to be used in bioenergy production is expected to play a multifunctional role including food production, source of energy and fodder, biodiversity conservation, and environmental services, as well as mitigation of the impact of climate change.

As described in Chapter 2, several limitations to the expansion of energy crops have been highlighted in recent years, particularly regarding the use of food crops for biofuels. At the same time, while biomass residues are considered for bioenergy, limitations to the use of forestry, industrial, and agricultural waste materials have been suggested to constrain the expansion of bioenergy projects. In Europe dedicated energy plantations would be required to meet the established goals regarding CO_2 emission reductions by 2020 and 2050 (European Environment Agency, 2007).

In tropical countries, the use of bagasse for pulp, ethanol, and cogeneration has strongly changed the business structure of sugarcane mills and ethanol plants, with a consequent biomass valorisation. More sources of raw materials are under development. Companies, governments, and stakeholders have already started to improve their knowledge on dedicated energy crops. Something similar has been occurring in Mediterranean countries where agricultural and industrial residues from cereal straw, forestry, olives, and vineyards are already being used for bioenergy. Private companies and governments have also started to evaluate risks of supply residues and often consider energy plantations as feedstock.

Until now, the promotion of dedicated energy crops has occurred with a limited consideration of the range of possible options. Most knowledge is centred on annual crops for first-generation biofuels with little attention to biomass costs and the environmental impact of many species already studied and used for other purposes (e.g., pulp, fibres, biochemicals). On the other hand, most trends suggest an increased interest on lignocellulosic crops and perennial species as feedstock for energy, and many countries in the world are already considering large investments in their development. Availability of resources is an important factor for high shares of biomass to penetrate electricity, heat, or liquid fuel markets. Lignocellulosic crops could promote perennial pastures and afforestation schemes in many countries that have an increasing demand for energy (and food). Their development as a source of biomass and energy requires a more sustainable pattern of land use, integrated with other forms of agriculture.

At present, more than two-thirds of global cropland is sown to monocultures of annual crops, and a large proportion of the most suitable lands for annual crops are already in use. This highlights the potential importance of perennial crops for biomass purposes. Perennial crops would address many agricultural problems, and may have substantial ecological and economic benefits relative to annual crop species; they can produce more ground cover and have longer growing seasons and more extensive root systems, which make them more competitive against weeds and more effective at capturing nutrients and water. They do not require annual tillage and consequently imply lower use of fuel-driven machinery. Therefore, perennial crops can be used in reducing soil erosion, minimising nutrient leaching, sequestering more carbon in soils, reducing the requirement of fuel for equipment, and providing continuous habitat for wildlife.

The interest in the use of tall grasses and perennial species as feedstock for solid biofuels has been increasing during the past two decades (Lewandowski et al., 2003). Many cool and warm annual grasses such as wheat, rye, and sorghum have been considered as feedstock for bioenergy purposes. The production and use of biomass for energy has both positive and negative impacts on the environment according to several national and European environmental organisations. Current evidence suggests that a conversion from annual to perennial crops is likely to have more positive environmental effects, particularly in relation to greenhouse gas (GHG) emissions and energy balance (Rowe et al., 2009; Fazio and Monti, 2011). In Spain two power energy plants of 16 MW and 25 MW consume hundreds of thousands of tonnes per year of cereal straw plus triticale, rye, and oats.[1] Some have suggested that most annual crops have higher costs and GHG emissions compared to perennial grasses even in low competitive lands when replacing natural gas with electricity (Maletta et al., 2012).

Most countries adopted some form of environmental assessment legislation in order to determine the implications of actions in advance, either at the project (Environmental Impact Assessment [EIA]) or at the strategic level for policies, plans, and programs (Strategic Environmental Assessment). In Europe bioenergy crops are considered for the replacement of traditional crops although European regulations prevent member states from reducing the area of permanent pasture (EC, 2003). Despite this, the future situation in Europe is less clear: projected long-term, contrasting scenarios accommodating both different socioeconomic conditions and climate scenarios indicate that a number of different outcomes are possible as soon as 2035 (European Environment Agency, 2007).

In a completely different situation, subtropical and tropical regions have increased planting of energy crops such as sugarcane (*Saccharum* spp.), a perennial crop which is argued to be replacing pasture land (i.e., natural perennial grass). It seems clear that successful policies will have to optimise in integrative ways the increasingly globalised (i.e., spatially separated) land-use and biomass utilisation chains. A major challenge to face for scenarios about the coming few decades (2050) will require the integration of perennial crops in marginal or low competitive lands considering biodiversity, linkages with food production, and renewable energies (Erb et al., 2012).

In this chapter, lignocellulosic energy crops with relevant updated information are analysed considering herbaceous and woody species. Nowadays, many energy crops have been investigated in the search of high productivity but not always considering low inputs, energy balances, or environmental impacts when their biomass is processed to energy. In the next sections, perennial energy canes and grasses are described in two main groups.

The so-called warm grasses are those species with four carbons (C4) in their photosynthetic metabolic pathway; they are more adapted to warmer climates such as those in the tropics and subtropics or parts of the Mediterranean basin. Species with three carbons (C3), also called cool grasses, are better suited for temperate climates; they are much less efficient in the use of water and nitrogen but can offer significant productivity in many regions, where C4 grasses are technically not feasible or where their costs and energy ratios are unreasonably high. Woody crops are also considered in short rotation coppice schemes for the production of lignocellulosic biomass feedstock.

13.2 Warm perennial grasses

Switchgrass

Switchgrass (*Panicum virgatum*) is a fast-growing, perennial, warm-season, C4 grass that occurs naturally from Canada to Central America. Several attributes make this species interesting as a bioenergy crop, including its good productivity, high environmental adaptability, suitability for marginal lands, low input requirements, and net energy yield (McLaughlin et al., 2002). Switchgrass is efficient in its use of nutrients; however, fertilisation is not recommended in the establishment year to avoid weed competition, which (with the seed dormancy typical of some warm seasons grasses) can increase planting fail (Sanderson et al., 2002).

Recommended nitrogen fertilisation is around 50 kg nitrogen per ha per year with just one harvest in the late autumn using common farm machinery (Lemus et al., 2008). Switchgrass cultivars are many. Recent research studies have selected new cultivars for energy purposes based on the survivability and productivity of switchgrass stand. Parrish and Fike (2005) found a strong correlation between latitude of origin and yield. Switchgrass can persist up to 10 years or longer; however, the crop takes three years to achieve its full yield potential (Schmer et al., 2008). Switchgrass has potential for use as a biofuel feedstock partially because it is productive under a wide range of conditions. Studies conducted in the USA and Europe showed encouraging results. In research plot, switchgrass has occasionally produced 22–33 tonnes of dry matter (DM)/ha per year (Elbersen et al., 2008). However, on a commercial scale it is probably more reasonable to expect 11–22 tonnes of DM/ha per year (Lemus et al., 2008).

Studies conducted in the EU and USA have shown that lowland switchgrass is more productive than upland cultivation at most locations, except at high northern latitudes. In the USA, annual yields average 12.9 tonnes of DM/ha per year for lowland and 8.7 tonnes of DM/ha per year for upland ecotypes, while with experimental and demonstrative projects, review of literature shows 0.9–34.6 tonnes of DM/ha per year (Lewandowski et al., 2003). Some field sites in Alabama, Texas, and Oklahoma reported biomass yields exceeding 28 tonnes of DM/ha per year using the lowland cultivars Kanlow and Alamo, the two most productive cultivars tested in the USA (Fuentes and Taliaferro, 2002; Sanderson et al., 2002).

In Europe switchgrass trials show a higher productivity in temperate areas with longer growing seasons (Mediterranean areas), compared with the northern European regions where reported productivity is lower, with frequent values of 6.5 tonnes of DM/ha per year.

However, the most important economic considerations for any energy crop are yield, production costs, competing purposes, and the price of competing feedstock (Mapemba et al., 2008). For this reason, maintaining high yield and keeping the cost low will result in the best economic returns in switchgrass production and management. Therefore, switchgrass generally achieved acceptable biomass yields, and its management required very little investment in terms of additional machinery or implements.

Regarding cultivar selection, there is still great uncertainty on whether lowland or upland ecotypes should be used in northern European countries. Biomass productivity is clearly the most important determinant in selecting energy crops in Europe. For this reason, the expectations for switchgrass as an energy crop are still significantly lower compared to other perennial and annual grasses that may outyield it: giant reed, sorghum, and Miscanthus in southern Europe and in northern Europe. The advantage of switchgrass compared to other competing perennial grasses mainly lies in the results of its integrated assessment; that is, by weighing all the operational, economic, and environmental aspects. Switchgrass is propagated by seed and requires very little investment in terms of farm machinery and agricultural inputs (Monti et al., 2009). Along with these important operational advantages, switchgrass could also provide significant environmental and economic benefits compared to other energy crops.

Considering bioethanol production in biorefineries, biomass from switchgrass can be broken down into sugars that can be fermented into ethanol

(Pimentel et al., 2005), with an estimated conversion efficiency of 380 litres of ethanol/tonne biomass feedstock (Morrow et al., 2006). Regarding non-North American and non-European countries in which switchgrass has been studied as an energy crop, a study conducted in 2005 by Parrish et al. reported that it is in use in 20 countries. Besides the North American and European nations already noted, countries that have produced studies on switchgrass include Argentina, Australia, China, Colombia, Japan, Korea, Mexico, Pakistan, Poland, Sudan, and Venezuela. In most cases, especially for the more recent citations, the studies deal with switchgrass for energy.

Miscanthus

Miscanthus, a genus of about 15 species of perennial grasses, is native to subtropical and tropical regions of Africa and southern Asia, with one species (*M. sinensis*) extending north into temperate eastern Asia (Heaton et al., 2010). The sterile hybrid between *M. sinensis* and *M. sacchariflorus*, *M. giganteus*, has been trailed as a biofuel in Europe since the early 1980s. Compared to Napier grass and other C4 species, Miscanthus can be established and produce acceptable yields even in cold environments, for example, in the UK, where 8,000 hectares were cultivated in 2011.[2] The rapid growth, low mineral content, and high biomass yield of Miscanthus make it a favourite choice as a biofuel.

In many studies, the majority of which have been done in Europe, Miscanthus has yielded more than 40 tonnes of DM/ha per year (Baldwing, 2008; Pyter et al., 2008). Miscanthus cultivation has been increasing steadily in the USA in the past few years, where comparisons are inevitably being made with switchgrass (Baldwing, 2008). A potential drawback of Miscanthus is that it must be propagated vegetatively via canes or rhizome cuttings, which complicates establishment. Typically full production can be expected by the second or third year (Lewandowski et al., 2003; Clifton-Brown et al., 2001). It is thought that replanting may be necessary after 15 years (Lewandowski et al., 2003). The costs of Miscanthus establishment can be a major limiting factor in areas where alternative crops can be directly sown as switchgrass or reed canary grass. Miscanthus establishment by rhizomes can cost from €1,900 to €2,800 per ha (Huisman et al., 1997; Khanna et al., 2008; Heaton et al., 2010).

Miscanthus can be used as input for ethanol production, often outperforming corn and other alternatives in terms of biomass produced. Additionally, it can be burned after harvest to produce heat and steam for power turbines. Also, when evaluating its carbon load and considering the amount of CO_2 emissions from burning the crop, any fossil fuels that might have been used in planting, fertilising, harvesting, and processing the crop – as well as in transporting the biofuel to the point of use – must also be considered. Its advantage, though, is that it is not usually consumed by humans, making it a more available crop for ethanol and biofuel than, say, corn and sugarcane.

Miscanthus is a crop that might be of interest not only in the EU and USA but also in many areas of tropical and subtropical countries. It has a high yield potential and a low fertiliser demand. Besides, established Miscanthus plants have no or only a very low pesticide requirement, and the leaching of nitrate from Miscanthus fields to the groundwater is very low.

Miscanthus can reach heights exceeding 3.5 metres in one growth season. Dry weight annual yield can reach 25 tonnes of DM/ha per year. For future application of Miscanthus within the energy sector, there will be a demand for rational techniques and systems for harvesting, handling, transporting, and final use. The field harvesting techniques must be in harmony with the techniques used subsequently by the end users. The harvesting may take place in the period from October to November when the crop has passed its ripening stage and until the following spring when the plants again begin to sprout. The combustion quality will be highly dependent on the time of harvesting. The crop's moisture content will decrease from 60–70% in the autumn to less than 20% in April, which is a clear advantage compared to other C4 grasses.

In terms of energy conversion, to meet the 2022 US biofuel target mandate, actions must be taken and biofuel crops (such as Miscanthus) instead of food crops (maize) must be cultivated just to avoid the food competition. Giant Miscanthus is capable of producing up to 20 tonnes of biomass and 12,000 litres of ethanol fuel annually per ha. Regarding ethanol production, harvest in May will represent a valid alternative. In such analyses, using maize to produce enough ethanol to offset 20% of US gasoline consumption would divert 25% of US cropland currently in production, while getting the same amount of ethanol from Miscanthus would divert 9.3%. If such projections can be proven in the marketplace, Miscanthus could help the USA to reach its target of replacing 30% of the gasoline it purposes with biofuels by 2030 (Pyter et al., 2008).

Napier grass

Napier grass (*Pennisetum purpureum* Schumach.) is also known as elephant grass. Napier grass used to be promoted in developing countries for soil conservation and for mulching coffee. The grass was then promoted mainly as a livestock feed. In recent years, the dwarf "Mott" Napier cultivar has been bred in Gainesville (Florida, USA) with a maximum height of about 1.5 m (Hanna and Monson, 1988). Unlike the tall variety, it is leafy and non-flowering. Napier grass is propagated vegetatively because seeds have low genetic stability and viability (Humphreys, 1994). Napier grass is a robust perennial forage with vigorous root system. Mature plants normally reach up to 4 metres in height and have up to 20 nodes. Boonman (1997) found it growing to a height of 10 metres in riverbeds and recorded a harvest of 29 tonnes of DM/ha per year taken in one cut on a very mature stand (more than two years overdue). Other authors reported yields from 40 to 49 tonnes of DM/ha per year (Keffer et al., 2009) in Hawaii. Our own experience year in Mexico and Brazil showed that large-scale projects should expect yields between 15 and 25 and from 25 to 35 tonnes of DM/ha per year unirrigated and irrigated areas, respectively; these figures correspond to locations with rainfall ranging from 1,400 to 1,800 mm per year and fertile soils suitable for sugarcane production.

Conventionally, napier grass is established in well-prepared land (ploughed and harrowed) from root splits, canes with three nodes, or whole canes. The material is planted 15–20 cm deep with splits planted upright. Three node canes are planted at an angle of 30–45° while whole canes are buried in the furrow 60–90 cm apart. Root splits generally take more labour to prepare (uproot) and to plant, but they result in quicker establishment and earlier, higher forage yields. Napier grass can be managed in areas suitable for sugarcane in the tropics, and there are some large-scale experiences in Latin America, Africa, and Asia for forage, pulp, and lastly for bioenergy. Large-scale projects have yields from 12 to 20 under rain-fed conditions and can easily achieve 40 tonnes of DM/ha per year with irrigation support (two or more cuttings).

The harvest usually takes place 220–280 days from cutting. Lignification of the plants can be improved through agronomy management in the dry season, where biomass moisture levels may decrease significantly (from 80% to 50%). Harvest management requires specific techniques to facilitate moisture losses (Lewandowski et al., 2003). Logistics require optimisation through management from the field to obtain lower costs. Harvest and moisture levels can be the most constraining factors if lignification is desired. If higher content of hemicellulose and soluble carbohydrates is required, several cuttings can be applied, and wet biomass handling will require specific logistic-reducing cost-optimisation measures.

Leaving appropriate stubble height will provide sufficient carbohydrate reserves for subsequent growth, and especially the stubble of the last harvest before the long dry period will encourage fast growth after the onset of rains. Lack of adequate and high-quality feed is a major constraint to production on smallholder farms, particularly in dry periods. The alternative is ensiling the surplus (Cuhna and Silva, 1977) because leaving napier grass to become too mature may compromise the quality. However, the proportion of farmers ensiling is small. Vegetative propagation is a major cost in large-scale projects, and crop lifetime is expected to be 7–10 years in most tropical countries.

13.3 Cool perennial grasses and other species

Lewandowski et al. (2003) concluded that 'bioenergy crops with C4 metabolic pathways produce higher yields compared to C3 species, but cooler environments would require C3 perennial species to produce similar yields with lower inputs.'

In temperate and warm regions, C4 grasses outyield C3 grasses due to their more efficient photosynthetic pathway. However, the further north perennial grasses are planted, the more likely cool-season grasses are to yield more than warm-season grasses. Low winter temperatures and short vegetation periods are major limits to the growth of C4 grasses in northern Europe. With increasing temperatures towards central and southern Europe, the productivity of C4 grasses and therefore their biomass yields and competitiveness increase.

When analysing potential C3 bioenergy crops, a critical factor is planting propagation systems (rhizomes and seeds) since productivity is not as high as C4 grasses in most cases and lower inputs are feasible for large lifetime periods in most cases. The main characteristics of four C3 and C4 perennial grasses can be found in Table 13.1.

Energy canes

Major cool energy canes with rhizomatous systems are being intensively studied and developed worldwide. Giant reeds (*Arundo donax*) and several species of bamboo canes are a matter of study for those searching for cheap and highly productive energy crops as feedstock for biomass-to-energy projects, mainly in the EU, the USA, Asia, and Latin America. Both species have a vegetative propagation establishment, and their productivity is a matter of discussion since most experiences with annually mechanised cuttings for biomass production have occurred in small plots or with irrigation support and in few regions; thus, uncertainty is becoming a limiting factor for the expansion of these species.

Bamboos include 1,250 species within 75 genera, most of which are relatively fast growing, attaining stand maturity within five years but flowering infrequently. Stands of tall species may attain heights of 15–20 metres, and the largest known (e.g., *Dendrocalamus giganteus*) grow up to 40 metres in height and 30 cm in culm (stem) diameter. Bamboo can be established from seeds, wildlings, or vegetative propagation.

Flowering of bamboo is an important factor to consider. Bamboo can flower after 40–80 years, and what triggers flowering intervals is not well understood. Seeds are infrequently available, and their use is not always a viable method for large-scale propagation. As well as with giant reeds, bamboo lifetime has not been described enough when culms are removed yearly in a bioenergy chain harvest system, and the information is vague and not comparable (Scurlock et al., 2000).

Giant reed is a sterile, high-energy perennial capable of producing more than 45 dry tonnes per ha. Along with this, the giant reed's fast-growing nature produces a bamboo-like stalk reaching upwards of 5 metres and allows for harvesting at least once per year after the establishment year. The giant reed's water uptake is lower during spring and summer and it can tolerate extreme drought events in the Mediterranean basin

Table 13.1 Characteristics, cultivation requirements and agri-environmental effects of four perennial grasses

	Miscanthus	Switchgrass	Reed Canary Grass	Giant Reed
Photosynthesis system	C4	C4	C3	C3
Height	Up to 4 m	Up to 2.5 m	Up to 2 m	Up to 5 m
Rotation time	15 years	15 years	10–15 years	15 years
Adaptation	Moderate winters, sufficient/low moisture	Moderate winters, sufficient/low moisture	Colder regions, moist conditions	Warm regions, moist conditions
Adaptation range in Europe	Cool and warm region of Europe (Denmark to Greece)	Cool and warm region of Europe (Denmark to Greece)	Cold and wet regions of northwest Europe (Scandinavia, UK, Netherlands, Eastern Europe)	Southern Europe, southern France, Italy, Greece, Spain
Propagation	Rhizomes	Seeds	Seeds	Rhizomes
Fertiliser input	In northern EU up to 50 kg N; in the south 50 to 100 kg N	In northern EU up to 50 kg N; in the south 50 to 100 kg N	Higher than for C4 grasses	Higher than for C4 grasses
Pesticide input	Low, possibly in first year	Low, possibly in first year	Low, possibly in first year	Low, possibly in first year
Runoff potential	Low	Low	Low	Low
Water use	Low	Low	Medium	Medium
Field pass frequency	1× per year after establishment	1× per year at harvest	1 time per year at harvest	1× per year at harvest
Erosion control	Good/Very good	Very good	Very good	Good/very good
% slope of terrain	Only machinery limitations need to be taken into account	Only machinery limitations need to be taken into account	Only machinery limitations need to be taken into account	Only machinery limitations need to be taken into account
Risk of fires	High	High	High	High

Source: Adapted from European Environment Agency (2007).

(Angelini et al., 2005). Harvesting through mowing and baling operations with existing machinery is possible, and is only problematic in tropical environments where stalks need to be dried (with any of several methods) during rainy seasons.

An important limitation of both species is the initial cost of establishment because labour and rhizome plantation machinery are necessary. Establishment from rhizomes in commercial projects requires proper machinery systems. There are many companies providing planting services in Europe, with costs of establishment from €2,000 to €3,000 per ha. Despite the fact that high yields can be achieved, as many companies providing planting material estimate, this economic cost scenario implies a high break-even point and cost recovery. This could imply very large risks when estimating the inputs and outputs of a large-scale project.

Another limiting factor of giant reeds and bamboo is that they are somehow difficult to eradicate and require farmers to have specialised expertise using mechanical and chemical treatments that take the ecology of the grasses, rhizome reserve accumulations, and regrowth into account, as well as the subsequent selection of crops and cultivation methods to avoid an invasive behaviour and undesired re-growth. In numerous countries, the giant reed has escaped cultivation and grows as a wild plant. It is a serious pest in the USA, Mexico, and South Africa. It is particularly problematic in California, where it has formed pure stands of more than thousands of ha of riparian habitat. These infestations have reduced natural biodiversity and pose a significant fire risk (Bell, 1997).

Grasses, grasslands, and C3/C4 mixtures

Most grasses have been investigated for forage production during the past few decades, producing huge quantities of information regarding management and agronomy as well as environmental benefits. The establishment of these crops from seeds requires a normal traditional seeder or broadcast sowing machine that can plant grasses in large areas with minimum constraints. Most grasses, as with cultivated switchgrass, admit non-tillage establishment, pasture renovation, and interseeding. This implies that a simple operation can renew the prairie and reestablish the plantation for another cycle with minimum emissions and low operation costs.

Several species can be considered possible C3 perennial species as feedstock for energy purposes. Nevertheless, some in particular are relevant to describe because more information is available and their potential is being considered in many regions of the world. Most C3 energy crops have frost tolerance during establishment, as well as higher yields in temperate climates and high altitudes compared to C4 species that require base temperatures higher than 12°C (e.g., sorghum and napier grass). Additionally, competition with annual weeds in semiarid regions usually limit the establishment of perennial C4 grasses.

Furthermore, C3 species have been mentioned in several studies as providing highly valuable sustainability criteria. This is the case of cardoon (*Cynara cardunculus*) in semiarid lands in southern European countries (Gominho et al., 2011; Ierna et al., 2012), reed canarygrass (*Phalaris arundinacea*) in the northern EU (Lewandowski et al., 2003), and tall wheatgrasses (*Elytrigia elongatum / Thinopyrum ponticum*) in several regions including North America and continental Mediterranean (Maletta et al., 2012) and eastern European countries (Csete et al., 2011). Another C3 species that seems to have similarly sound scientific evidence regarding productivity and sustainability potential in temperate climates is Virginia fanpetals (*Sida hermaphrodita*) in Poland (Borkowska and Molas, 2013).

Promoting a more sustainable bioenergy chain from bioenergy crops involves considering the environmental benefits shown in most cases by these grasses compared to rhizomatous crops and species with high inputs and requirements, even with much higher yields. In this regard, many C3 grasses occupying native stands and natural prairies in temperate regions in the EU and USA have been evaluated as feedstock for bioenergy with promising results. Some examples of the latter have been addressed when analysing grasslands in large extensions such as those of North America, Europe, Russia, Argentina, or Australia, where significant amounts of land could shift to the production of perennial crops if a large market for bioenergy and bio-based products emerges.

De La Torre Ugarte et al. (2003) and McLaughlin et al. (2002) indicated that this could happen if prices for energy crops were high enough to attract farmers interest: 'At a farm gate price of US \$44 per tonne of DM, an agricultural sector model predicts higher profits for switchgrass than conventional crops on 16.9 million hectares (ha). Benefits would include an annual increase of US\$6 billion in net farm returns, a US\$1.86 billion reduction in government subsidies, and displacement of 44–159 Tg/year (1 Tg = 10^{12} g) of greenhouse gas emissions' (McLaughlin et al., 2002). Natural grasslands and native species could be eventually improved and managed to produce more biomass through several techniques that have been already described in traditional forage research. These practices include intercropping of genetic improved species, C4 and C3 grass mixtures, and even legumes to reduce fossil energy inputs of fertiliser application (that can be more than 70% of total inputs of energy outputs when considering the life-cycle assessment).

Other traditional practices such as pasture renovation through reestablishment techniques with direct sowing machinery are frequent and can be considered to avoid high costs and fossil inputs when renovating perennial grasses' lifetimes. This would mean lower inputs per production unit in the long run.

Regarding carbon sequestration, several studies have reported recently large benefits from perennial C3 bioenergy crops and grasslands. More than 13 million ha of former cropland are enrolled in the US Conservation

Reserve Program, providing well-recognised biodiversity, water quality, and carbon sequestration benefits that could be lost on conversion back to agricultural production.

A recent study provided measurements of the GHG consequences of converting grasslands to continuous corn, corn–soybean, or perennial grass for biofuel production. Projected carbon debt repayment periods under no-till management range from 29 to 40 years for corn–soybean and continuous corn, respectively. Under conventional tillage, repayment periods are three times longer, from 89 to 123 years, respectively. Alternatively, the direct use of existing grasslands for cellulosic feedstock production would avoid carbon debt entirely and provide modest climate change mitigation immediately (Gelfand et al., 2013).

Other authors have suggested that winter cereals used for lignocellulosic production as feedstock could have higher production costs and lower emission savings when replacing fossil energy compared to perennial C3 grasses in semiarid lands even with yields lower than five tonnes per ha per year (Maletta et al., 2012).

Mixtures of a warm-cool season of perennial grasses could be possible in the same stand both in natural prairies or cultivated pastures. In the USA biomass yields for the warm-season usually range from 4.0 to 7.2 tonnes of DM/ha per year without any inputs (e.g., fertilisers) and 3.4–6.0 tonnes of DM/ha per year for cool-season grasslands. Higher environmental benefits from grasslands and mixtures compared to other biofuel sources have been suggested and highlighted in several studies including McLaughlin and Walsh (1998), Lewandowski et al. (2003), and Gelfand et al. (2013).

13.4 Short rotation coppice

Short rotation coppice (SRC) is a plantation characterised by a very short rotation: between two and four years. Species grown in SRC are mainly willow and poplar, but black locust may also be used. Willow is grown mainly in the northern parts of the EU.

A plantation could be viable for up to 30 years before replanting becomes necessary, although this depends on the productivity of the stools. Willow or poplar will grow rapidly in the first year reaching up to 4 metres in height. During the winter, after planting, the stems are cut back to ground level to encourage the growth of multiple stems (i.e., coppiced). The crop is harvested generally three years after cutback and again during the winter. The equipment used for harvesting is specifically developed for the purpose and depends on the fuel specification of the customer or end user. Most operations other than planting or harvesting can be completed using conventional farm machinery. In the UK yields from willow SRC at first harvest are expected to be in the range of 7–12 tonnes of DM/ha per year depending on site and efficiency of establishment, while yields from poplar are expected to be 12–20 tonnes of DM/ha per year.

Table 13.2 Main characteristics of SRC

	Willow	Poplar	Black Locust
Location	Northern, central, and western Europe	Central Europe	Mediterranean Europe
Crop density stools/ha	18–25,000	10–15,000	8–12,000
Rotation years	1–3	3–4	2–4
Average butt diameter at harvest (mm)	15–30	20–50	20–40
Average height at harvest (m)	3.5–5.0	2.5–7.5	2.0–5.0
Growing stock at harvest (fresh tons/ha)	30–60	20–45	15–40
Moisture content (% weight)	50–55	50–55	40–45

Source: EUBIA (2007).

Current studies show that most of the crops and trees can be cultivated across different regions of Europe (see Table 13.2). In each region or country, the yield and input parameters are different and therefore the price of resulting biomass will differ.

For some countries, where SRF (Short Rotation Forestry) or SRF have been cultivated for some years, production costs are available, but there are differences in the way countries calculate their costs. The total costs can be separated into direct costs (consumable inputs) and labour/machine costs, both specified as costs in Euro per ha. Direct costs regard consumable inputs such as seeds, pesticides, and fertilisers. Labour and machine costs do not have the same meaning in the different countries. Sometimes the labour costs are explicitly mentioned (e.g., Germany), or the labour costs are integrated into the costs for different activities.

Examples of labour and machine costs are costs for ploughing, using the power harrow, rotary cultivating, drilling, rolling, spraying, fertilising, irrigating, mowing, baling, and transporting. Little can be said about current trading structures since in most countries there seems to be no market for biomass for SRF or SRC. For the establishment of short rotation coppice in agricultural systems, the mechanisation of the whole process is fundamental. Only planting and harvesting of the trees, depending on the method, can take from 20% to 60% of the total costs. The harvesting costs are dominating, since crops are harvested up to 15 times (Scholz, 2007).

13.5 Conclusion

Companies and governments interested on feedstock development and production to feed power and biofuel facilities face several challenges regarding lignocellulosic energy crops. As we have discussed in this chapter, many uncertainties still require attention before starting a biomass-to-energy project based on dedicated energy plantations.

Lignocellulosic perennial crops have enough strong evidence to be considered as feedstock and realistic sustainable alternatives for bioenergy. In particular, more work is needed for the development of alternatives in marginal and less competitive areas, which will necessarily have low aerial productivity. Lower inputs at the farm level and highly efficient processing technology will be relevant to determine sustainable production patterns in those low-yielding regions.

The large-scale production of biomass from perennials and their introduction in rotation patterns can be more profitable for farmers compared to annual rotation and monocultures (Hallam et al., 2001). The introduction of energy plantations in arable and fallow lands may enhance yields in the subsequent cropping system as rotations with perennials often build up organic matter while avoiding soil erosion. Several sources showed that introducing perennial systems in farmlands can determine a higher biodiversity and soil improvements with predictable and measurable effects for farmers (Pedroli et al., 2012; Shulte et al., 2013).

There is a lack of reliable information and demonstration projects that could be used as baseline studies when planning a biorefinery in a new region. Yield estimations and production cost analyses for specific locations are a key issue for developers and promoters of bioenergy chains.

Agronomic models and information on the optimisation of the several possible species to consider need further research and development programs to reduce uncertainty and provide reliable data for estimations on costs and environmental impacts. Logistic optimisation is required in many energy plantations in tropical areas where biomass drying could be a serious limiting factor. The existence of various species and high variability in inputs and site performances require expertise and specific management to reduce impacts and costs. Residue supply and biomass from lignocellulosic crops complement each other and could help stakeholders in achieving a sustainable integrated pattern with food production.

Another important limitation is the fact that private companies are providing information on their own research studies and breeding while offering services on those specific developments, with scarce independent advice on other alternatives. This results in a serious risk for farmers being misled by companies selling planting material with a business plan advertising their own estimation of expected yields and costs. Even when some crops have been only tried in small experimental plots and have not been studied long enough (and on a wide enough range of conditions), verifiable information and productivity estimations from demonstration projects sometimes are critical.

Farmers require technology and governmental extension services since they are still reluctant to adopt lignocellulosic perennial crops. Long-term contracts for supply to biorefineries are a key issue. Increased demand for biomass for bioenergy purposes may lead to a continued conversion of valuable habitats into productive lands and to intensification, which both have negative effects on biodiversity. Whether cropping and harvesting biomass have positive or negative effects will depend strongly on specific site conditions, reference systems replaced, and processing technology for a specific productivity. Evaluating biodiversity and other specific environmental impacts of bioenergy crops requires selecting low-input species. Lignocellulosic perennial herbaceous and woody species should be encouraged over annual cropping systems, particularly in marginal areas; degraded lands; and less competitive, more fragile environments.

Notes

1. More information is available at http://www.acciona-energia.com/.
2. More information on current factsheets of Miscanthus cultivation in the UK can be found on the NNFCC website: http://www.nnfcc.co.uk.

References

Angelini, L. G., Ceccarini, L., and Bonari, E. (2005). Biomass yield and energy balance of giant reed (*Arundo donax* L.) cropped in central Italy as related to different management practices. *European Journal of Agronomy,* Vol 22, pp375–389.

Baldwing, B. (2008). *Cultured feedstocks.* Sustainable Energy Research Center (SERC), Mississippi State University.

Bell, G. (1997). Ecology and management of *Arundo donax,* and approaches to riparian habitat restoration in Southern California. In J.H. Brock, M. Wade, P. Pysek, and D. Green (Eds.), *Plant invasions: Studies from North America and Europe* (pp. 103–113). Leiden, the Netherlands: Blackhuys.

Borkowska, H. and Molas, R. (2013). Yield comparison of four lignocellulosic perennial energy crop species. *Biomass and Bioenergy*, 51, pp.145–15

Borkowska, H., and Molas, R. (2013). Yield comparison of four lignocellulosic perennial energy crop species. *Biomass and Bioenergy.* ISSN 0961–9534.

Clifton-Brown, J. C., Lewandowski, I., Andersson, B., Basch, G., and Christian, D. G. (2001). Performance of 15 Miscanthus genotypes at five sites in Europe. *Agronomy Journal,* Vol 93, pp1013–1019.

Csete, S., Stranczinger, S., Szalontai, B., Farkas, A., Pál, R. W., Salamon-Albert, E., Kocsis, M., Tóvári, P., Vojtela, T., Dezsö, J., Walcz, I., Janowszky, Z., Janowszky, J., and Borhidi, A. (2011). Tall wheatgrass Cultivar Szarvasi-1 (*Elymus elongates* subsp. *ponticus* cv. Szarvasi-1) as a potential energy crop for semi-arid lands of Eastern Europe, sustainable growth and applications in renewable energy sources. M. Nayeripour (Ed.), InTech, DOI: 10.5772/26790. Retrieved on 10 November 20130 from: http://bit.ly/148ck2h

Cuhna, P. G., and Silva, D. J. (1977). Napier grass silage without concentrate supplementation as the only feed for beef cattle in dry season. *Cientifica,* Vol 5, pp65–69.

De La Torre Ugarte, D. G., Walsh, M. E., Shapouri, H., and Slinsky, S. P. (2003). *The economic impacts of bioenergy crop production on U.S. agriculture.* U.S. Department of Agriculture, Agricultural Economic Report No. 816.

EC. (2003). Council Regulation (EC) No 1782/2003 of 29 September 2003 establishing common rules for direct support schemes under the common agricultural policy and establishing certain support schemes for farmers and amending Regulations. *Official Journal of the European Communities*, L270:1e70.

Elbersen, H., Bindraban, P., Blaauw, R., and Jongman, R. (2008). *Biodiesel from Brazil*. The Hague: Ministry of Agriculture, Nature and Food Quality.

Erb, K.-H., Haberl, H., and Plutzar, C. (2012). Dependency of global primary bioenergy crop potentials in 2050 on food systems, yields, biodiversity conservation and political stability. *Energy Policy*, Vol 47, pp260–269.

EUBIA (European Biomass Industry Association). (2007). *Biomass procurement: Other energy crops*. Retrieved from http://www.eubia.org/index.php/about-biomass/biomass-procurement/other-energy-crops

European Environment Agency (EEA). (2007). *Land-use scenarios for Europe: Qualitative and quantitative analysis on a European scale*. Copenhagen: European Environment Agency.

Fazio, S., and Monti, A. (2011). Life cycle assessment of different bioenergy production systems including perennial and annual crops. *Biomass and Bioenergy*, Vol 35, pp4868–4878.

Fuentes, R. G., and Taliaferro, C. M. (2002). Biomass yield stability of switchgrass cultivars. In J. Janick and A. Whipkey (Eds.), *Trends in new crops and new uses* (pp. 276–282). Alexandria, VA: ASHS Press.

Gelfand, I., Sahajpal, R., Zhang, X., Izaurralde, R. C., Gross, K. L., and Robertson, G. P. (2013). Sustainable bioenergy production from marginal lands in the US Midwest. *Nature*, Vol 493, pp514–517.

Gominho, J., Lourenço, A., Palma, P., Lourenço, M. E., Curt, M. D., Fernández, J., and Pereira, H. (2011). Large scale cultivation of *Cynara cardunculus* L. for biomass production: A case study. *Industrial Crops and Products*, Vol 33, pp1–6.

Hallam, A., Anderson, I. C., and Buxton, D. R. (2001). Comparative economic analysis of perennial, annual, and intercrops for biomass production. *Biomass and Bioenergy*, Vol 21, pp407–424.

Hanna, W. W., and Monson, W. (1988). Registration of dwarf Napier grass germplasm. *Crop Science*, Vol 28, pp870–871.

Heaton, E. A., Dohleman, F. G., Miguez, A. F., Juvik, J. A., Lozovaya, V., Widholm, J., Zabotina, O. A., McIsaac, G. F., David, M. B., Voigt, T. B., Boersma, N. N., and Long, S. P. (2010). Miscanthus: A promising biomass crop. In J.-C. Kader and M. Delseny (Eds.), *Advances in botanical research* (pp. 75–137). London, UK: Academic

Huisman, W., Venturi, P., and Molenaar, J. (1997). Costs of supply chains of Miscanthus giganteus. *Industrial crops and products*, Vol 6, pp353–366.

Humphreys, L. R. (1994). *Tropical forages: Their role in sustainable agriculture*. Harlow, UK: Longman.

Ierna, A., Mauro, R. P., and Mauromicale, G. (2012). Biomass, grain and energy yield in Cynara cardunculus L. as affected by fertilization, genotype and harvest time. *Biomass and Bioenergy*, Vol 36, pp404–410.

Keffer, V. I., Turn, S. Q., Kinoshita C. M., and Evans, D. E. (2009). Ethanol technical potential in Hawaii based on sugarcane, banagrass, Eucalyptus, and Leucaena. *Biomass and Bioenergy*, Vol 33, pp247–254.

Khanna, M., Dhungana, B., and Clifton-Brown, J. (2008). Costs of producing Miscanthus and switchgrass for bioenergy in Illinois. *Biomass and Bioenergy*, Vol 32, pp482–493.

Lemus, R., Brummer, E. C., Burras, C. L., Moore, K. J., Barker M. F., and Molstad N. E. (2008). Effects of nitrogen fertilization on biomass yield and quality in large fields of established switchgrass in southern Iowa, USA. *Biomass and Bioenergy,* Vol 32, pp1187–1194.

Lewandowski, I., Scurlock, J.M.O., Lindvall, E., and Christou, M. (2003). The development and current status of perennial rhizomatous grasses as energy crops in the US and Europe. *Biomass and Bioenergy,* Vol 25, pp335–361.

Maletta, E., Martín, C., Ciria, P., del Val, M. A., Salvado, A., Rovira, L., Díez, R., Serra, J., González Y., and Carrasco, J. E. (2012). *Perennial energy crops for semiarid lands in the Mediterranean: Elytrigia elongata, a C3 grass with summer dormancy to produce electricity in constraint environments.* 20th European Biomass Conference and Exhibition, Milan, June 18–22.

Mapemba, L. D., Epplin, F. M., Huhnke, R. L., and Taliaferro, C. M. (2008). Herbaceous plant biomass harvest and delivery cost with harvest segmented by month and number of harvest machines endogenously determined. *Biomass and Bioenergy,* Vol 32, pp1016–27.

McLaughlin, S. B., De La Torre-Ugarte, D. G., Garten Jr., C. T., Lynd, L. R., Sanderson, M. A., Tolbert, V. R., and Wolf, D. D. (2002). High-value renewable energy from prairie grasses. *Environmental Science and Technology,* Vol 36, pp2122–2199.

McLaughlin, S. B., and Walsh, M. E. (1998). Evaluating environmental consequences of producing herbaceous crops for bioenergy. *Biomass and Bioenergy,* Vol 14, pp317–324.

Monti, A., Fazio, S., and Venturi, G. (2009). Cradle-to-farm gate life cycle assessment in perennial energy crops. *European Journal of Agronomy,* Vol 31, pp77–84.

Morrow, W. R., Griffin, W. M., and Matthews, H. S. (2006). Modeling switchgrass derived cellulosic ethanol distribution in the United States. *Environmental Science and Technology,* Vol 40, pp2877–2286.

Parrish, D. J., and Fike, J. H. (2005). The biology and agronomy of switchgrass for biofuels. *Critical Reviews in Plant Sciences,* Vol 24, pp423–459.

Pedroli, B., Elbersen, B., Frederiksen, P., Grandin, U., Heikkilä, R., Henning Krogh, P., Izakovičová, Z., Johansen, A., Meiresonne, L., and Spijker, J. (2012). Is energy cropping in Europe compatible with biodiversity? Opportunities and threats to biodiversity from land-based production of biomass for bioenergy purposes. *Biomass and Bioenergy,* 10.1016/j.biombioe.2012.09.054, Vol 55, pp. 73–86

Pimentel, D., Hepperly P., and Hanson, J. (2005). Environmental, energetic and economic comparisons of organic and conventional farming systems. *Bioscience,* Vol 55, pp573–582.

Pyter, R., Voigt, T., Heaton, E., Dohleman, F., and Long, S. (2008). *Growing giant Miscanthus in Illinois.* University of Illinois. ISO 690. Retrieved from http://miscanthus .illinois.edu/wp-content/uploads/growersguide.pdf

Rowe, R. L., Street, N. R., and Taylor, G. (2009). Identifying potential environmental impacts of large-scale deployment of dedicated bioenergy crops in the UK. *Renewable and Sustainable Energy Reviews,* Vol 13, pp271–290.

Sanderson, M. A., Adler, P., Skinner, R. H., Dell, C., and Curran, B. (2002). *Establishment, production, and management needs of switchgrass for biomass feedstock in the Northeastern USA.* Chapel Hill, NC: The North Carolina Botanical Garden.

Schulte, L.A., Todd, Ontl, A. and Drake Larsen, G.L. (2013). Biofuels and Biodiversity, Wildlife Habitat Restoration. In: *Encyclopedia of Biodiversity* (Sec. ed.), S.A. Levin (ed.), pp. 540–551. Waltham: Academic Press.

Schmer, M.R., Vogel, K.P., Mitchell, R.B., and Perrin, R.K. (2008). Net energy of cellulosic ethanol from switchgrass. *Proceedings of the National Academy of Sciences,* Vol 105, pp464–469.

Scholz, V. (2007). Mechanisation of SRC production. *Bornimer Agrartechnische Berichte,* Vol 61, Potsdam-Bornim, Germany.

Schulte, L.A., Todd, Ontl, A. and Drake Larsen, G.L. (2013). Biofuels and Biodiversity, Wildlife Habitat Restoration. In: *Encyclopedia of Biodiversity* (Sec. ed.), S.A. Levin (ed.), Pp. 540-551. Waltham: Academic Press.

Scurlock, J.M.O., Dayton, D.C., and Hames, B. (2000). Bamboo: An overlooked biomass resource? *Biomass and Bioenergy,* Vol 19, pp229–244.

14 Bioenergy production from waste agricultural biomass

P.M.F. Quist-Wessel and J.W.A. Langeveld

14.1 Introduction

The development of bioenergy offers major possibilities for the reduction of greenhouse gas (GHG) emissions and fossil fuel dependency. It may also, however, cause unintended negative impacts; for example, it can affect existing land-use patterns, food production, or biodiversity. This offers a dilemma for policy makers, who need to determine how to promote sustainable ways of bioenergy development to replace fossil fuel use without jeopardising other policy objectives. Many studies have identified agricultural and industrial biomass residues as promising feedstocks that bring fewer risks with respect to competition for food or affecting natural resources. The amount of residues available for energy production, the way in which they should be converted, and the form emerging bioenergy chains can take remains to be determined.

An inventory of biomass availability presented by Dornburg et al. (2010) suggests that agricultural and forestry residues could provide an equivalent of 85 EJ of energy in 2050. According to IEA Bioenergy (2011), wastes and residues together with sustainably grown energy crops should be able to provide the majority of the biomass feedstock requirements needed to realise biofuel, as well as heat and power targets in 2050, but the exact amount and character remain unrevealed.

The term *waste agricultural biomass* (WAB) refers to those organic materials that do not directly go into foods or other products but are necessarily generated during crop production or processing. Mostly, this biomass is in the form of residual stalks from crops, leaves, roots, seeds, seed shells, and so on. It is estimated that globally, approximately 5 billion metric tonnes of agricultural waste is generated every year – the thermal equivalent of approximately 1.2 billion tonnes of oil – about 25% of the current global production (UNEP, 2012). Converting WAB into energy has environmental as well as economic benefits. WAB is a clean source of energy, as the carbon cycle loop is closed (the carbon dioxide released by combustion is again sequestered in the next crop), and usually there are no harmful emissions.

The lack of clear, robust, and long-term policy framework, together with the need to incorporate new and often complex technologies, can be a major challenge for the development of new production chains. Uncertain economic conditions (including lack of capital), limitations in know-how and institutional capacities, underdeveloped biomass and carbon markets, problems in chain coordination, and limited public support have been identified as the main barriers for bioenergy chains development (Langeveld et al., 2010).

One major activity within the framework of IEA Bioenergy is to develop strategies to integrate expanding bioenergy systems with existing land-use patterns in order to reduce risks for land-use competition and displacement. The aim is to help to mobilise bioenergy feedstock applications, improving land-use productivity and reducing negative environmental impacts of the existing land use. Much emphasis is put on facilitating informed and balanced decision making and supporting this process by providing scientific data and analysis.

The development of sound bioenergy options requires balanced and solid empirical data on bioenergy production and its impact on land use and local inhabitants, while procedures are required that can use these data in an evaluation process to guide decision making based on a wide set of considerations (Langeveld et al., 2013). This chapter uses data and methods developed in the framework of IEA Bioenergy Tasks 30 and 43 – alongside other work – in an attempt to quantify the availability and potential use of biomass residual streams.

The main focus is on biogas production from waste agricultural biomass. First, information is presented on processes involved in biogas production (Section 14.2). Biogas production from WAB is illustrated with two case studies: (1) use of waste from green coffee processing in Honduras (Section 14.3) and (2) use of palm oil residues in Colombia (Section 14.4). The chapter ends with a brief discussion.

14.2 Biogas production

Biogas is produced in a process of anaerobic fermentation in which organic material is converted by microorganisms into methane (CH_4) and carbon dioxide (CO_2) under oxygen-free conditions. The overall anaerobic digestion process can be depicted as follows:

$$\text{Organic matter} \rightarrow CH_4 + CO_2 + \text{water} + \text{minerals} + \text{microbial biomass} + \text{organic residue} \qquad (14.1)$$

Methane and carbon dioxide together form the biogas. The major minerals produced are ammonium, phosphate salts, and hydrogen sulphide. The mineral solution including the organic residue is referred to as *digestate* and is an effective organic fertiliser.

Digestion and biogas production

The process of anaerobic digestion of organic material is presented in Figure 14.1, which depicts four major steps that can be distinguished in the fermentation process:

1. *Hydrolysis* – conversion of polymers into monomers (sugars, fatty acids, and amino acids)
2. *Acidogenesis* – conversions of monomers into volatile fatty acids (VFAs), alcohols, hydrogen gas, ammonia, and carbon dioxide
3. *Acetogenesis* – conversion of VFAs and alcohols into acetate, hydrogen, and carbon dioxide
4. *Methanogenesis* – conversion of acetate, hydrogen, and carbon dioxide into methane

Each step is conducted by a specific group of anaerobic bacteria. These groups operate synergistically, reinforcing each other's efficiency. The final performance of the fermentation process thus depends on the accumulated performance of different groups of bacteria, each with its own requirements and sensitivities. This makes management of the process complex, requiring constant monitoring of process conditions that include temperature, acidity, substrate feeding, and biomass composition (e.g., C:N ratio). The main process characteristics are depicted in Table 14.1.

The final efficiency of biogas production can reach 80–95% (Angenent and Wrenn, 2008). However, because anaerobic digestion is a sensitive

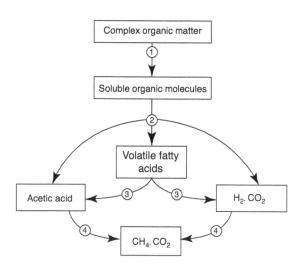

Figure 14.1 Anaerobic pathway of digestion of organic material
Source: Zupančič and Grilc (2012).

Table 14.1 Environmental requirements for anaerobic digestion

Parameter	Hydrolysis/Acidogenesis	Methanogenesis
Temperature	25–35°C	Mesophilic: 30–40°C Thermophilic: 50–60°C
pH value	5.2–6.3	6.7–7.5
C:N ratio	10–45	20–30
C:N:P:S ratio	500:15:5:3	600:15:5:3
Trace elements	No special requirements	Essential: Ni, Co, Mo, Se

Source: Zupančič and Grilc (2012).

process, process failures might occur when a group of micro-organisms is inhibited or – alternatively – overloaded.

Mesophilic anaerobic fermentation, which requires moderate temperatures (30–40°C) and combines stable fermentation conditions with high yields, is the most commonly used process. However, yields of mesophilic aerobic fermentation are lower in comparison to thermophilic fermentation (50–60°C). Optimal pH values for hydrolysing and acidifying bacteria range between 5.2 and 6.3; acetogenic bacteria require a pH of 6.8–7.5 (Zupančič and Grilc, 2012). Single-unit processes preferably should have a pH of 7.5 (Zwart and Langeveld, 2010).

Most organic materials (including crops, crop residues, manure, and industrial residues) are suited for anaerobic digestion. Productive feedstock generally contains 15–20% dry matter, is high in volatile solids (VS), contains fat, is relatively high in protein, and is low in lignin (Zwart and Langeveld, 2010). The required C:N ratio is between 10 and 30 (Zupančič and Grilc, 2012); at higher ratios carbon is not optimally converted into methane.

Feedstock retention time is defined as the average time spent by feedstock in the fermenter and is calculated by dividing fermenter content by the average daily feedstock load. Ideal retention times are 25–40 days for mesophylic reactions and 15–25 days for thermophylic reactions (Zwart and Langeveld, 2010).

Biogas application

Biogas contains methane (CH_4), carbon dioxide (CO_2), and water (H_2O), plus relatively small amounts of ammonia (NH_3) and hydrogen sulphide (H_2S). Methane concentration ranges from 45% to 70%, but usually from 50% to 55%. Biogas yields can be expressed in different units such as kg (or m^3) of biogas (methane) produced per kg dry matter, per kg VS or per kg feedstock (Zwart and Langeveld, 2010). Waste can generate about 250–500 m^3 biogas per tonne of dry matter (Yu et al., 2010).

Table 14.2 Replacement values of biogas

	Biogas Required for Replacement (m³)
1 kg of firewood	0.2
1 kg of charcoal	0.5
1 meal cooked for 1 person	0.15–0.3
Boiling 1 litre of water	0.03–0.04
1 lamp used for a day	0.12–0.15
Generate 1 kWh of electricity with a generator	1

Source: UNESCAP (2007).

There are many applications for biogas, including household and industrial production of heat, light, or electricity. For small-scale production units, it can be connected to kitchen stoves or used for gas lamps. Biogas with a methane content of 60% has a caloric value of 21.4 KJ/L (Rodriquez and Zambrano, 2010), and 1 m³ corresponds with 6 kWh or 0.5–0.6 litre of diesel fuel. Table 14.2 provides some data on the use of biogas in replacing common energy sources. Biogas from industrial anaerobic digesters can be converted into electric power using combined heat and power (CHP) installations, or it can be injected into the gas grid.

Digestate

The composition of the digestate depends on the feedstock used. It is considered a good fertiliser because it contains high quantities of organic matter and nutrients and is odourless. On average, 0.03–0.15 kg of dry matter is produced per kg chemical oxygen demand (COD) fed to the digester (Wilkie, 2008). According to Reinhard (2006), the majority of nutrients are retrieved in the digestate (86% of nitrogen and 70% of phosphate). This makes it an excellent fertiliser that typically contains 7.5% of solids (Table 14.3).

Digestate is often applied as a mix of inorganic (ammonium and phosphate) and organic fertiliser, although application is subject to local legislation

Table 14.3 Typical composition of digestate

	Unit	Quantity
Total solids (TS)	%	7.5
Volatile solids (VS)	%	5.8
Carbon	kg/ton TS	417
Nitrogen	kg/ton TS	19
Phosphorus	kg/ton TS	4.3

Source: Van Buren (1979).

(Zwart and Langeveld, 2010). Practices of digestate application in existing crop production systems are currently being developed; for example, in main biogas production regions of northwest Europe where this type of fermentation is actively supported by bioenergy policies. However, much remains to be done in fitting biogas production and digestate application in existing production practices that are subject to increasingly strict environmental policies (Langeveld et al., 2013).

14.3 Coffee waste in Honduras

Coffee is one of the most important cash crops worldwide, providing incomes to large numbers of smallholders in Latin America (Brazil, Colombia, Honduras), Africa (Ethiopia), and Asia (Vietnam, Indonesia). It is cultivated in about 70 countries covering nearly 10.5 million ha, with an average annual production (2009–2011) of about 8.4 million tonnes of green or unroasted coffee (FAOSTAT, 2013).

With an annual production of more than 259,000 tonnes of Arabica green coffee (2009–2011), Honduras ranks sixth in global coffee production (FAOSTAT, 2013). More than 282,000 ha were cultivated in 2011, with an average yield of 950 kg green coffee per ha (FAOSTAT, 2013). Nearly all coffee production (95%) is realised by small producers who have less than two ha.

Coffee processing

The Arabica coffee cherries are processed by wet processing, which means that water is used for removal of the outer parts of the fruit (i.e., skin [exocarp], pulp [mesocarp], and the mucilage layer). The coffee is dried before the parchment is removed and what remains is 'green coffee'. The outturn of clean dry coffee of 100 kg red-ripe cherries is 18.2% for Arabica coffee (Wintgens, 2009).

Processing is shown in Figure 14.2 and consists of the following steps:

- Reception (rinsing) and initial sorting
- Depulping: Ripe fruits are pulped 12–24 hours after harvesting through mechanical separation of pulp from the beans. The water used for this step is referred to as pulping water
- Fermentation and washing: The de-pulped seeds then ferment in holding tanks to degrade the mucilage and allow the residual fruit juice to drain out. This juice is referred to as sweet water. Through washing, the mucilage is removed (waste water)
- Drying: After washing, the moisture content of the wet parchment coffee is about 52%, and the seeds are dried (11–12% moisture content) either by sun-drying, or through a suitable mechanical dryer, or by a combination of both
- Shelling and grading: The parchment is removed and the remaining green coffee is graded and ready for export

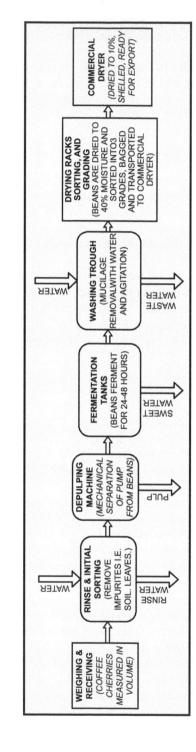

Figure 14.2 Scheme of coffee processing (wet method)

Source: Ferrell and Cockerill (2012).

Table 14.4 Overview of typical residual flows during processing of 5.5 tonnes Arabica berries

Waste Stream	Moisture Content (%)	Amount (tons)	Percentage of Fruit Weight (%)	Bioenergy Potential
Fresh pulp	82	2.25	41	Biogas/ (bioethanol)
Mucilage	65–75	0.9	16.3	Bioethanol/ biogas
Husks	10	0.25	4.5	Heat, power
Water content wet parchment		1.1	20	
Processing water		22.0		Biogas

Source: Wintgens (2009).

By-products from coffee processing include pulp, mucilage, husks, and processing water. The processing of 5.5 tonnes of berries generates 1 tonne of green coffee, 2.2 tonnes of pulp, 900 kg of mucilage, and 250 kg of parchment. Average water use for the production of 1 tonne of green coffee varies between 40 and 80 m^3 in systems without circulation of process water and between 8 and 22 m^3 in systems with recirculation of water (Wintgens, 2009). Based on observations of Ferrell and Cockerill (2012), wastewater production is estimated at 108 litres per tonne of green coffee. Table 14.4 provides an overview of the by-products of green coffee processing and their bioenergy potential.

Coffee processing waste is often dumped into the waterways of the upper watersheds without control or treatment. The concentration of organic matter disposed into these streams is sufficient to cause eutrophication, with the following subsequent issues: (1) strong odours and attraction of flies and other insects, (2) waste water with a high oxygen demand that has the effect of depleting the water of oxygen and consequently killing aquatic life, (3) pollution of local water sources, and (4) GHG emissions (CH_4, N_2O) from pulp and waste water basins (Steiner, 2011).

Husks that have a caloric value of 17.9 MJ per kg (Rodriquez and Zambrano, 2010) are used as fuel in burners for heat exchangers of coffee driers.

Biogas production and yield

Coffee pulp, mucilage, and wastewater are rich in organic matter, which makes them suitable feedstocks for biogas production. The expected energy yield depends on their composition, conversion technology, and efficiency of the different production processes. Fermenting coffee waste is done using linear plug flow systems, which have equal input and output volumes.

Introduced high nutrient and energy load feedstock are pushing out the effluent with low energy contents (Ferrell and Cockerill, 2012).

The design depends on the size of the digester and materials available for construction. Low-cost, bag-style digesters are most common in Latin America and utilise a plastic membrane over a trench or lagoon with an earth, concrete, or plastic-lined bottom. The plastic membrane builds positive pressure as it expands with gas, which can be used directly in a stove or boiler application. This system, however, has limited gas storage capacity (Ferrell and Cockerill, 2012). Companies that design, build, and provide technical assistance currently also offer more sophisticated and industrial designs. According to von Enden (2002), the required investments of between US$6,000 and US$10,000 can only be gained for the large-scale operation of at least 80 tonnes of fresh fruits per day.

Data on biogas yields for fresh pulp vary in the literature from 25 m^3/tonne (Rodriquez and Zambrano, 2010) to 60 m^3/tonne (Steiner, 2011) and 71 m^3/tonne (GTZ, 2010) with a methane content of 63%. For further calculations, we used an average value of 52 m^3 per tonne of fresh pulp. According to Steiner (2011), biogas production for coffee wastewater is 1.3 m^3 biogas per m^3 of wastewater with a removal efficiency of 80%.

Extrapolating these figures to the national coffee production of Honduras with an annual pulp production of 570,000 tonnes and a water use of 5.7 million m^3 results in a biogas potential of 29.6 million m^3 and 7.4 million m^3 biogas for pulp and water respectively.

14.4 Palm oil residues in Colombia

The area dedicated to oil palm cultivation in Colombia has more than doubled in the past decade and reached 402,000 ha in 2010. A further increase of around 150,000 ha has been foreseen by the government (Gamba et al., 2011). In 2004 and 2005 national average yield was 4.1 tonnes of crude palm oil per ha (Fedepalma, 2011), but yields dropped mainly due to diseases and climatic problems, and average yield in 2010 was 3.0 tonnes of oil per ha (Gamba et al., 2011). Crude palm oil production amounted to 753,000 tonnes in 2010 (FAOSTAT, 2013), of which 100,000 tonnes was exported (Gamba et al., 2011).

Fresh fruit bunches (FFB) and loose fruits are milled within 24 hours after harvesting to preserve palm oil quality. The scheme of palm oil processing is presented in Figure 14.3. The FFB are sterilised by steaming under high pressure, and the fruitlets are separated from the palm bunches. The crude palm oil is obtained through mechanical pressing. By means of centrifugation, the crude palm oil (CPO) is separated from waste and water.

The production of one tonne of CPO requires five tonnes of FFB. On average, the processing of one tonne FFB in palm oil mills generates 230 kg empty fruit bunches (EFB) and 650 kg palm oil mill effluent (POME) (Stichnothe and Schuchardt, 2010).

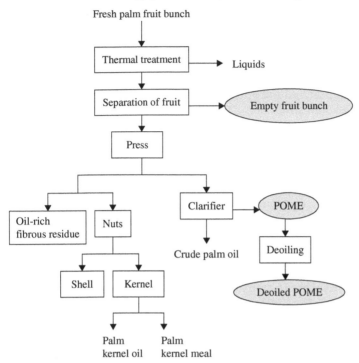

Fresh palm fruit bunch

Thermal treatment → Liquids

Separation of fruit → Empty fruit bunch

Press

Oil-rich fibrous residue | Nuts

Clarifier → POME

Crude palm oil

Shell | Kernel

Deoiling

Deoiled POME

Palm kernel oil | Palm kernel meal

Figure 14.3 Palm oil processing
Source: Fang (2010).

EFB can be used as compost feedstock, but are often dumped. POME is often treated in pond systems with land application or discharge of the outlet. But the high methane emissions from pond systems, up to 8 m³/tonne of FFB are a serious problem for the atmosphere, and discharge of POME to surface water is not only a pollution of the environment but also a waste of nutrients (Stichnothe and Schuchardt, 2010). A practical way to capture biogas is the use of a fermenter system.

Biogas production of POME in fix-bed fermenters yields 8.9 m³ per tonne of FFB with a methane content of 63% (Stichnothe and Schuchardt, 2010). Based on the Colombian crude oil production of 753,000 tonnes in 2010, Colombia has the potential to produce 33.5 million m³ of biogas from POME.

14.5 Discussion and conclusion

Crop and industrial biomass waste have been heralded as sustainable sources of bioenergy that have little or no risk for environmental damage or negative social implications. In practice, however, inventories of waste availability and quality remain very general, and suggestions as to how effective, safe, and economic production chains based on waste valorisation should be developed are scarce. When done properly, the production of biogas from waste agricultural biomass can contribute directly and indirectly to GHG emission reduction by avoiding methane generation from both agricultural waste and processing wastewater. Additionally, it provides a carbon-neutral alternative to fossil fuels. Other environmental benefits are the avoidance of water and air pollution and the option to return nutrients to the plantation through digestate application.

The size of the digester will depend on the availability of WAB, the investment capability, and the application or market for the energy. The coffee harvest in Honduras is from early November to the end of March, implying that additional feedstock sources are required during the rest of the year. In case of oil palm, the feedstock is available through the year as FFB are harvested year round with a FFB peak occurring in March–April (Henson et al., 2011).

There is a large range in the size of coffee processing units, and digester size and technology will have to be selected accordingly. Biogas produced by small digesters (daily biogas production of 8 m^3) will most likely be used for household cooking and to replace firewood, which is most common in the coffee-growing areas of Honduras. It will reduce the workload, mainly for women in firewood collection, and improve the indoor air quality by replacing firewood. If biogas is used for light, it might allow children and adults to study in the evening. In this way, the technology can contribute to regional development.

Larger coffee processing units and oil palm production units can install industrial digesters. The energy that is produced (in the form of biogas and – some – additional heat) can be integrated in the processing plant. The benefits are (1) cost reduction – by replacement of energy that would have to be purchased otherwise – and (2) clean production – through replacement of a fossil fuel source.

Waste management significantly influences the environmental performance of processing units and adds to the green image of the sector. Specifically for organic coffee producers this is an asset that could translate into higher prices. The challenge to be met is to develop sound, appropriate technology that does not interfere with the delicate coffee processing. This will require adequate research and training programs oriented towards staff of coffee plants. Active involvement of staff and other stakeholders will be a prerequisite for successful implementation of the digesters.

References

Angenent, L.T., and Wrenn, B.A. (2008). Optimizing mixed-culture bioprocessing to convert wastes into bioenergy. In J.D. Wall, C.S. Harwood, and A. Demain (Eds.), *Bioenergy* (pp. 179–194). Washington, DC: ASM Press.

Dornburg, V., van Vuuren, D., van de Ven, G., Langeveld, H., Meeusen, M., Banse, M., van Oorschot, M., Ros, J., van den Born, G.-J., Aiking, H., Londo, M.,

Mozaffarian, H., Verweij, P., Lysen, E., and Faaij, A. (2010). Bioenergy revisited: Key factors in global potentials of bioenergy. *Energy and Environmental Science,* Vol 3, pp258–267.

Fang, C. (2010). Biogas production from food-processing industrial wastes by anaerobic digestion. PhD thesis, DTU Environment, Lyngby, Denmark.

FAOSTAT. (2013). Retrieved 22 February 2013 from http://faostat3.fao.org/home/index.html#HOME

Fedepalma. (2011). *Oil palm information system.* Retrieved 25 February 2013 from http://portal.fedepalma.org//statistics.shtm

Ferrell, J., and Cockerill, K. (2012). Closing coffee production loops with waste ethanol in Matagalpa. Nicaragua. *Energy for Sustainable Development,* Vol 1, pp44–50.

Gamba, L., Peters, D., and Toop, G. (2011). *Sustainable biomass identification mission for Colombia: Stakeholder and sector analysis.* Utrecht, the Netherlands: Ecofys.

GTZ. (2010). *Agro-industrial biogas in Kenya: Potentials, estimates for tariffs, policy and business recommendations.* Retrieved from http://www.scribd.com/doc/36890933/Biogas-Kenya

Henson, I. E., Romero, R. R., and Romero, H. M. (2011). The growth of the oil palm industry in Colombia. *Journal of Oil Palm Research,* Vol 23, pp1121–1128.

IEA Bioenergy. (2011). *Technology roadmap: Biofuels for transport.* Paris, France: International Energy Agency.

Langeveld, J.W.A., Kalf, R., and Elbersen, H. W. (2010). Bioenergy production chain development in the Netherlands: Key factors for success. *Biofuels, Bioproducts and Biorefining,* Vol 4, pp484–493.

Langeveld, J.W.A., Quist-Wessel, P.M.F., van Esch, J.W.A., and Berndes, G. (2013). Bioenergy alongside other land use: sustainability assessment of alternative bioenergy development scenarios. *Journal of Forest Energy,* Vol 23, pp48–63.

Reinhard, B. (2006). *Valoriation concepts for biodigestor effluents.* Zürich, Switzerland: HSW University of Applied Science. Retrieved from http://www.empa.ch/plugin/template/empa/*/59249/---/l=2

Rodriquez, V., and Zambrano, F. (2010). *Los subproductos del café: Fuente de energía renovable.* Bogotá, Colombia: Federacion Nacional de Cafeteros de Colombia.

Steiner, R. (2011). *Energetic use of residues from coffee production in Central and South America.* Presentation held at the workshop Energetic Use of Residues from Coffee Production in Central and South America - 21st of June 2011, Zollikon, Ursen, Switzerland: Renewable Energy & Energy Efficiency Promotion in International Cooperation (REPIC).

Stichnothe, H., and Schuchardt, F. (2010). Comparison of different treatment options for palm oil production waste on a life cycle basis. *International Journal of Life Cycle Assessment,* Vol 15, pp907–915.

UNEP. (2012). *Background paper on converting waste agriculture biomass into energy.* Paper presented at the Biennium Conference of the Global Partnership on Waste Management, Osaka, Japan, 5 and 6 November 2012. Osaka: Global Partnership on Waste Management.

UNESCAP. (2007). *Recent developments in biogas technology for poverty reduction and sustainable development.* United Nations Economic and Social Commission for Asia and the Pacific (UNESCAP). Asian and Pacific Centre for Agricultural Engineering and Machinery (UNAPCAEM), Beijing, China. p. 76.

Van Buren, A. (1979). *A Chinese biogas manual: Popularising technology in the countryside*. London: Intermediate Technology Publications.

von Enden, J. C. (2002). Re-use of processing residues, post harvest processing Arabica coffee, GTZ. Retrieved from http://www.venden.de/pdfs/Reuse_of_processing_residuesV2.pdf

Wilkie, A. C. (2008). Biomethane from biomass, biowaste and biofuels. In J. D. Wall, C. S. Harwood, and A. Demain (Eds.), *Bioenergy* (pp. 195–207). Washington, DC: ASM Press.

Wintgens, J. N. (2009). *Coffee: Growing, processing, sustainable production. A guidebook for growers, processors, traders and researchers*. London: Wiley-VCH.

Yu, Z., Morrison, M., and Schanbacher, F. I. (2010). Production and utilization of methane biogas a renewable fuel. In A. A. Vertès, N. Qureshi, H. P. Blaschek, and H. Yukawa (Eds.), *Biomass to biofuels: Strategies for global industries* (pp. 403–434). Chichester, UK: Wiley.

Zupančič, G. D., and Grilc V. (2012). Anaerobic treatment and biogas production from organic waste, management of organic waste. In S. Kumar (Ed.), *InTech*. Retrieved 12 December 2012 from http://www.intechopen.com/books/management-of-organic-waste/anaerobic-treatment-and-biogas-productionfrom-organic-wastes

Zwart, K. B., and Langeveld, J.W.A. (2010). Biogas. In J.W.A. Langeveld et al. (Eds.), *The biobased economy: Biofuels, materials and chemicals in the post-oil era* (pp. 180–198). London: Earthscan.

15 Impact on land and biomass availability

*J.W.A. Langeveld, J. Dixon and
H. van Keulen*

15.1 Introduction

Now that biofuel production in different major producing nations has been analysed, the current chapter brings the results of the previous chapters together. The main focus is on the amount of biofuels produced, the amount of land needed, and the impact for food availability. This will give a global assessment of the way biofuel production affects land use and food production.

The chapter is organised as follows. Global production of biofuels is presented in Section 15.2. Land-use requirements are reported in Section 15.3, and biomass use is discussed in Section 15.4, which is followed by a discussion on the relation between food availability and undernutrition in Section 15.5 and a brief conclusion in Section 15.6.

15.2 Biofuel production

In only a few years dynamic biofuel industries emerged in different countries around the world. The speed with which industrial activities have developed is amazing. In the USA a considerable number of biofuel factories were installed within a period of two or three years. Similar patterns have been observed, be it not so strong, in the EU and also in other countries, where biofuel industries have developed very quickly. To put this in context, most other agricultural food and fibre production industries developed over decades and sometimes centuries.

Chain and production capacity development are reflected by strong increases in fuel output. In 2010 the countries studied in this book produced 86 billion litres of ethanol and 15 billion litres of biodiesel, up from 17 and 1 billion, respectively, at the turn of the century (Table 15.1). Together, Brazil, the USA, the EU, Indonesia, Malaysia, Mozambique, South Africa, and China represent the vast majority of global biofuel production – more than 95% of the world's production of ethanol and three-quarters of the biodiesel. The balance of biodiesel production is generated in a range of countries, for example, Argentina (2 billion litres in 2010) and Canada (EIA, 2013).

Table 15.1 Biofuel production in 2000 and 2010 (billion litres)

	Ethanol			Biodiesel		
	2000	*2010*	*Increase*	*2000*	*2010*	*Increase*
Brazil	9.7	27.6	17.9	Neg	2.1	2.1
USA	6.1	49.5	43.4	Neg	2.1	2.1
EU	1.5	6.4	4.9	0.8	10.3	9.5
Indonesia/Malaysia	NI	NI	NI	Neg	0.2	0.2
China	Neg	2.1	2.1	Neg	0.4	0.4
Mozambique	Neg	0.02	0.02	Neg	0.05	0.05
South Africa	Neg	0.02	0.02	Neg	0.05	0.05
All	17.3	85.6	68.3	0.8	15.1	14.3

Billion = thousand million; NI = not included; Neg = negligible.

The USA is the largest producer of ethanol, generating nearly 50 billion litres, or more than half of world total. About half of the global supply of biodiesel is produced in the EU, followed by Brazil and the USA.

15.3 Land use

Data on land use for the expanding biofuel industry are presented in Table 15.2. Gross biofuel area quadrupled, reaching 31.5 million ha in 2010. This represents an increase of 24.9 million ha; 11.4 million ha of this is related to the production of co-products. Net biofuel land occupation increased by 13.5 million ha. Nearly half of this is found in the USA. Expansion in Brazil (4.9 million ha) and the EU (6.6 million) has been more modest.

Table 15.2 Land use for biofuel production 2000–2010 (million ha)

	Biofuel Crop Harvested Area 2000	*Biofuel Crop Harvested Area 2010*	*Increased Harvested Area*	*Associated with Coproducts*	*Net Increase Biofuel Area*
Brazil	2.6	7.5	4.9	1.8	3.1
USA	2.2	13.2	11.0	5.9	5.1
EU	2.1	8.7	6.6	3.2	3.4
Indonesia/ Malaysia	0.03	0.05	0.02	0.01	0.01
China	0	2.2	2.2	0.4	1.8
Mozambique	0	0.1	0.1	0.03	0.1
South Africa	0	0.1	0.1	0.04	0.1
All	7.0	31.5	24.9	11.4	13.5

Land use for biofuels in 2006, reported by Cotula et al. (2008), amounted to 14 million ha, which seems in line with our findings. This represents approximately 1% of global arable land. At the global level, biofuel production in 2030 might occupy 35–54 million ha of land (Cotula et al., 2008). This is 20–85% more than the area occupied in 2010. The European Biomass Association, AEBIOM, assessments suggest that 4–6 million ha of arable land in the EU was used for biofuel feedstocks in 2011. Our assessment is higher but includes co-product area, assuming a 50% co-product area gives a net biofuel area just exceeding 4 million ha, which coincides with AEBIOM assessments.

In order to put net increase of biofuel feedstock area in perspective, we also present other changes in land use that have been observed. Two types of changes are discussed here: (1) loss of agricultural area and (2) increased multiple cropping index (i.e., MCI). Since 2000 the countries included in our analysis lost 9 million ha of agricultural land (Table 15.3). This is the net outcome of area expansion in emerging and developing countries (Brazil, Indonesia/Malaysia, Mozambique; total amounting to 22 million ha) and of loss of agricultural area elsewhere (USA, EU, China, South Africa; total of 31 million ha).

The loss represents three-quarters of net biofuel area expansion. During the same period, however, intensification of land use generated additional harvest, amounting to 42 million ha. In other words, in response to the increased demand for biomass (for food, feed, and fuels), farmers have

Table 15.3 Biofuel expansion and other land-use changes between 2000 and 2010 (million ha)

	Increased harvested area	Associated with Co-products	Net Increase Biofuel Area	Changes in Agricultural Area	Released from Increased MCI	Change in NHA
Brazil	4.9	1.8	3.1	12.0	4.9	13.8
USA	11.0	5.9	5.1	–3.5	10.9	2.3
EU	6.6	3.2	3.4	–11.5	3.6	–11.2
Indonesia/ Malaysia	0.02	0.01	0.01	8.9	2.0	10.9
China	2.2	0.4	1.8	–13.4	20.3	5.1
Mozambique	0.1	0.03	0.1	1.3	0.9	2.0
South Africa	0.1	0.04	0.1	–2.7	–1.2	–4.0
All	24.9	11.4	13.5	–9.0	41.5	19.0
Global total				–47.8	91.5	

Sources: Biofuel expansion data taken from Chapters 6–12. Changes in agricultural land, harvested area ratio, and harvested food area calculated from FAOSTAT (2010–2013), http://faostat.fao.org and Chapters 6–12.

intensified cropping. Increases in MCI of the study area thus effectively gen-erated enough harvested area to compensate for biofuel area expansion, while also compensating losses of agricultural land. This is not the case for all countries included in the study. In the EU, loss of agricultural area is too high while South Africa demonstrates a decline in MCI. In most cases, how-ever, net area available for food crops (net harvested area) has increased. This is a remarkable feat that has major implications.

Two important conclusions can be drawn here. First, while biofuel area expanded, this did *not* lead to a net decline of land available for traditional food and feed markets. The net biofuel area expansion has been many times smaller than global increase in harvested area due to MCI increases, which amounted to more than 90 million ha of harvested crops in 2010. Second, biofuels are not threatening the availability of agricultural area. While area for agriculture is declining in many countries, the losses are mostly caused by other processes, including urbanisation, expansion of nature (locally), and touristic land use, as well as land degradation.

On an aggregated level, only 22% of area reduction is associated with the expansion of biofuel production. For individual countries, this can be higher. It is 23% for the EU and 60% for the USA. Calculations for Brazil and the Far East are not possible, because agricultural area has expanded in contrast to industrialised countries where it declined. Emerging countries (except for China and South Africa) show a net increase of food production area. The increase in MCI in China has been sufficient to compensate for the loss of agricultural area. In the USA and EU intensification was insufficient to compensate for the aggregated effect of increased biofuel production and loss of agricultural area.

These findings are surprising and do not confirm projections published elsewhere. Many studies concluded that biofuels are causing a decline in agricultural area. A World Bank report (Stage et al., 2009), for example, concluded that biofuel expansion would reduce availability of arable land and would also cause significant land reallocation and decreases in forest and pasture lands. Similar conclusions have been drawn by many others.

However, the data presented earlier show that biofuels are not the main cause of the decline in agricultural area. This means that they cannot be the main drivers leading to deforestation or grassland conversion, but data on loss of agricultural land due to urbanisation and so forth are difficult to obtain. Land losses reported for China (2.8 million ha over a period of two decades; as reported by Stage et al., 2009) seem to be modest.

Figures for the USA suggest an increase of urban and industrial land with 22 million ha during four decades (1954–1997). Since 1992, however, urbanisation in the USA seems to have accelerated, claiming 6.5 million ha of cropland between 1992 and 1997 alone. This is more than net biofuel area expansion in the USA in a *decade*. These figures are impressive. They should be compared to data on deforestation. Forest loss in tropical areas was estimated by Mongabay at 10 million ha during a period of 15 years, while illegal logging was estimated at some 1.4 million ha per year by Lawson and MacFaul (2010).

The dominant role of urbanisation in loss of agricultural land does not mean that expansion of biofuel area is – or cannot be – associated with deforestation. According to Macedo et al. (2012), between 2006 and 2010, 78% of the soybean production increase was realised through expansion of cultivated area. The remainder was due to yield increases. More than 90% of the expansion was on previously cleared forest land. Cropland expansion, however, did fall from 10% to 2% of deforestation, with pasture expansion accounting for most remaining deforestation. The authors found little evidence of leakage of soybean expansion into other regions (e.g., Cerrado in Mato Grosso), although indirect land-use changes could not be excluded (Macedo et al., 2012).

15.4 Biomass use

In 2010 considerable amounts of biomass were used in biofuel production (Table 15.4). Total biomass use for biofuel production more than doubled from 193 million tonnes to 517 million tonnes in 2010. More than 90%

Table 15.4 Biomass used in biofuel production in 2000 and 2010 (million tonne)

Biofuel Feedstock	Total Biomass Use			Co-products Generated	Net Use for Biofuels
	2000	*2010*	*2010–2000*	*2010–2000*	*2010–2000*
Cane in Brazil	170	348	178.0	0.0	178.0
Soybean in Brazil	Neg	9	9.3	5.4	3.9
Corn in USA	16	119	103.0	54.9	48.1
Soybean in USA	No data	5	5.0	2.9	2.1
Wheat in EU	Neg	7	7.0	3.7	3.3
Rapeseed in EU	6	20	14.0	6.5	7.5
Sugar beet in EU	No data	No data	10.0	4.0	6.0
Oil palm in Far East	0.5	1.0	0.5	0.1	0.4
Sugarcane in Mozambique	Neg	0.3	0.3	0.0	0.3
Sugarcane in South Africa	Neg	0.2	0.2	0.0	0.2
Wheat in China	0.0	1.7	1.7	0.9	0.8
Corn in China	0.0	4.2	4.2	2.2	2.0
Total in this book	192.5	516.9	324.4	78.2	246.2

Neg = negligible.

Source: Chapters 6–12.

of all feedstock is used for ethanol production, which is dominated by two crops: sugarcane (in Brazil) and corn (in the USA). Brazilian cane represents three-quarters of all biofuel feedstock (down from 90% in 2000). Cane ethanol production in other countries is almost negligible. In the EU the use of wheat amounted to 7 million tonnes only.

Biodiesel feedstocks are dominated by rapeseed in the EU, which represents close to 60% of all biodiesel production. The remainder is mostly made up by soybean in Brazil and the USA. Total biodiesel feedstock use is 38 million tonnes in 2010. The share of co-products increased from 6% in 2000 to 18% in 2010. This is due to the expansion of corn ethanol. While cane generates no co-products of significance (other than bagasse used for the generation of electricity), corn produces more than half of its weight in co-products. This depends, however, on the type of milling that is applied (wet milling provides more co-products than dry milling).

Stage et al. (2009) found that expansion of biofuel production (to meet national targets) would cause a reduction in biomass availability. Biofuels would therefore cause significant land reallocation and decreases in forest and pasture lands, as well as a reduction in food supply. The magnitude of the impact on food supply – at the global level not as large as perceived earlier – would be significant in developing countries such as India and countries in sub-Saharan Africa.

Many authors have presented similar conclusions. Often, the way biofuels affect food availability is through the way they cause food prices to rise (e.g., see Charles, 2009).

The impact of biofuel production on food crop availability is determined in Table 15.5, which compares increased net biomass use between 2000 and

Table 15.5 Biomass used in biofuel production in 2000 and 2010 (million tonne)

Biofuel Feedstock	Net Use for Biofuels	Changes in Biomass Availability	Increase of Biomass Productivity	Net Change in Biomass Availability
Brazil	181.9	465.5	478.4	283.6
USA	50.2	48.1	137.8	−2.1
EU	20.0	−30.5	111.8	−50.5
Indonesia/Malaysia	0.3	48.0	27.9	47.7
Mozambique	0.3	4.0	6.3	4.0
South Africa	0.3	−3.8	Negligible	−4.1
China[1]	3.4	342.4	412.0	339.0
Total in this book	253.7	873.8	1172.2	617.6
Global		1413.1	1434.8	

[1]Including biodiesel feedstocks.

Sources: Calculated from FAOSTAT (2010–2013), http://faostat.fao.org; Chapters 6–12.

2010 with changes in biomass availability during the same period and calculates net changes in biomass availability for food and animal feed purposes. Since 2000 nearly 900 million tonnes of extra biomass have become available. Most of this is found in Brazil, where biomass production expanded by 465 million tonnes. This means that one country alone has been able to produce sufficient additional crops to cover biofuel biomass use. Biomass production also increased in China (342 million tonnes), the USA (48 million tonnes), Indonesia/Malaysia, and Mozambique. Reduced production was reported in the EU and – especially – South Africa. Global biomass production has increased with 1.4 billion tonnes, more than four times biofuel feedstock use.

Table 15.5 also presents data on biomass productivity (BP), or the amount of biomass produced per ha of arable land. BP is responsible for most of the production increase, especially in Brazil, China, the USA, and the EU. It declined in South Africa. Often, increases in biomass availability from enhanced BP are larger than the net increase in biomass availability. The difference is the production capacity that has been lost due to the loss of agricultural land (e.g., USA, EU, China).

Changes in biofuel feedstock use and total biomass availability were used to calculate net change in biomass availability. Since 2000 crops available for food and feed markets increased by more than 600 million tonnes. Most of this was realised in China and Brazil. Food availability has risen in all countries with the exception of the USA, the EU, and South Africa. The largest loss is found in the EU, where 11.5 million ha of agricultural land was lost between 2000 and 2010.

15.5 Food availability and undernutrition

To better understand the relation between food crop production and undernutrition, this section discusses the prevalence of undernourishment in the study area. Data in Table 15.6 clearly show that food availability per head of the population between 2000 and 2010 has increased. This was the case in Africa, Asia, and Latin America, as well as in developed countries. Consequently, the number of undernourished has declined. The largest improvement was realised in Asia, where 64 million now have sufficient food to cover their needs. The largest relative reduction was observed in Latin America, which saw one in five undernourished significantly improve their situation. Africa is the only continent where the number of people with hunger has increased.

It is important to notice that *average* dietary supply is more than enough to cover the needs of the population. Total and regional food availability is adequate to provide sufficient food and is increasing. Between 2000 and 2010 average dietary energy supply increased by three percentage points to 120% of requirements. FAO confirms that the world produces enough food to feed everyone, with agriculture worldwide generating 17% more

Table 15.6 Global food availability and undernutrition

	Average Dietary Energy Supply (% of required)		Number Undernourished (million)		Decline Number Undernourished	
	2000	2010	2000	2010	Million	% of 2000
Africa	110	115	203	230	–27	–13%
Asia	112	116	603[1]	540[1]	64[1]	11%[1]
Latin America	122	126	46	37	9	19%
Developed	133	134	60[2]	56[2]	4[2]	7%[2]
Total	117	120	913	863	50	5%

[1]Including Oceania.
[2]Estimation.

Sources: Calculated from FAO (2013); Hunger Notes (2013).

calories per person today than it did 30 years ago – despite a 70% population increase (Hunger Notes, 2013).

While there is enough food to provide everyone in the world with 2,790 kilocalories (kcal) per day (FAO, 2011), in 45 countries one-third or more of the population is hungry. Most of these countries are in Africa. Four countries (Burundi, Eritrea, Ethiopia, and Zambia) see more than 40% of the people suffer from undernutrition. Together, these countries are home to more than 60 million underfed (Hunger Notes, 2013). These are tragedies, which are rightly brought to our attention.

Less than half the total cereal production – amounting to 2.2 billion tonnes – is actually used as food. One-third is used as animal feed. On the whole, less than 40% of cereals, pulses, and oil crops is used for food. Almost 800 million tonnes is used in livestock production. Nearly 120 million tonnes is applied in non-food and non-feed applications such as industrial use and biofuels.

The amount of food that is wasted is especially striking. FAO (2013) suggests that about 100 million tonnes of cereals are wasted each year. Per-capita waste by consumers in Europe, North America, and Oceania has been estimated at about 95–115 kg a year, while consumers in sub-Saharan Africa, South Asia, and Southeast Asia throw away 6–11 kg a year. Total waste exceeds 1 billion tonne per year.

The current estimate of underfed people is lower than estimates published immediately after food prices peaked in 2007–2008. In 2010 FAO claimed that 925 million people were undernourished. This would have indicated a net increase in the prevalence of hunger.

There was some confusion as to the main reasons behind the – alleged – increase. Several explanations were suggested: (1) neglect of agriculture relevant to the poor by governments and international agencies, (2) the economic crisis, and (3) the increase of food prices. Thus, a number of reports published since 2008 that were based on alleged rises in the number of hungry have proved to be incorrect. It is important to note that the general image of biofuels has been largely affected by this error.

Earlier we saw that the increased implementation of biofuel policies in biofuel-producing countries such as Brazil, the USA, and the EU did not affect biomass production or food availability in general, except through international price effects.

While there is sufficient *average* availability of food, in 2010 some 200 million hungry were reported in countries covered in this book (Table 15.7). Consequently, these countries represent one-quarter of all undernourished. Most of them are found in China, the country with low biofuel production. Improvements in food conditions have been strongest in Indonesia, another country that does not host significant biofuel production. Both countries showed a strong economic growth during the past decade.

In the study countries, prevalence of undernutrition declined by 25% in the period 2000–2010 (from 250 to 199 million). These data do not include undernourishment in South Africa, Malaysia, the USA, and the EU, where prevalence is set below 5% but not further quantified. The highest shares of undernourished people are in Mozambique (39%) and China (12%). The largest absolute number of undernourished is found in China (over 150 million people in 2010).

The reduction of undernourished in countries included in the book is four times higher than the global average, which is (at least partly) explained by their high economic growth. Further, data shown previously suggest no clear relation between economic growth, biofuel production, and decline (or lack of decline) of undernutrition. This is best demonstrated by comparing the

Table 15.7 Food availability and undernutrition in biofuel-producing countries

	Average Dietary Energy Supply (% of required)		Number Undernourished (million)		Decline Number Undernourished	
	2000	*2010*	*2000*	*2010*	*Million*	*% of 2000*
Brazil	121	132	21	14	7	32%
Indonesia	108	121	38	23	15	40%
Malaysia	126	124	–[1]	–[1]	No data	No data
Mozambique	94	100	8	9	–0.5	–6%
South Africa	121	126	–[1]	–[1]	No data	No data
China	117	123	182	153	29	16%
Total	–	–	250	200	51	20%

[1]Less than 5% of the population.

Source: Calculated from FAO (2013).

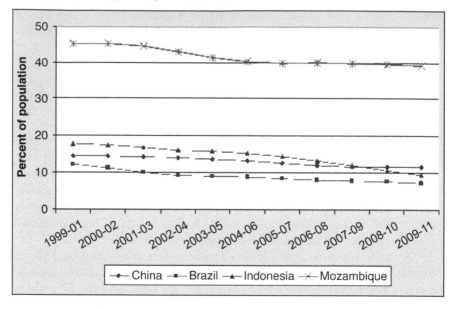

Figure 15.1 Changes in prevalence of undernutrition in Brazil, Indonesia, Mozambique, and China

Source: Calculated from FAO (2013).

situation in Brazil, Indonesia, and China. While the largest decline in hunger is reported in China (29 million people less in 10 years), in relative terms, Indonesia and Brazil have done better, diminishing the number of hungry in 2010 with 40% and 32%, respectively, as compared to 2000 (Figure 15.1).

From this we can make a few cautious observations. First, at the national level, high-level production of biofuels can coincide with a strong reduction in hunger (which was demonstrated in Brazil). Apparently it requires dedicated policies to reduce the level of (poverty and) undernutrition (Indonesia, Brazil). This is in line with analyses reported elsewhere. According to Wegner and Zwart (2011), hunger and malnutrition are due less to the unavailability of food and more to the inability of the poorest to access food at an affordable price. In addressing hunger, policy matters more than geography or history.

The exact relation between biofuel production and poverty and undernutrition is complex and depends strongly on local conditions, but biofuel production does not necessarily aggravate human problems. According to Cotula et al. (2008), biofuels can be instrumental in revitalising land use and livelihoods in rural areas. Price signals to small-scale farmers could significantly increase yields and incomes and secure long-term poverty reduction in countries that depend heavily on production of agricultural commodities for their income. Large-scale biofuel cultivation could also provide benefits in the form of employment, skills development, and secondary industry. The authors warn, however, that the legal position of smallholders to land should be guaranteed.

One element that is easily overseen is the role of co-products. The production of ethanol from corn and wheat, for example, does not draw too heavily on the nutritional value of the feedstocks because the co-product (distillers dried grains with solubles) can be used as livestock feed (Golub and Hertel, 2012).

15.6 Conclusion

While biofuel production has strongly increased throughout the past decade, the amount of land used to produce biofuel feedstocks has been limited. Almost half of the area can be associated with co-products, which leaves a total of 14 million ha for net biofuel production. An additional 9 million ha of agricultural area is lost to processes such as urbanisation, tourism, and land abandonment. Increasing MCI, however, has released sufficient land to compensate for the sources of land loss, and net land availability for food production has increased by 14 million ha.

These findings are in contrast to common beliefs that biofuel production comes at the expense of land and food availability. It is not clear how processes such as urbanisation and intensification (e.g., shortening of fallow or increasing double cropping) are covered in the analysis of biofuel expansion. In general, more attention should be given to make and compare complete land balances because these can be instructive in explaining the dynamics of land use in a given region. This should also include non-biofuel-producing countries such as the Russian Federation and the Ukraine. These are important food producers that seem to have lost some of their appetite for producing crops for export markets. While increasing demand of cereals in the USA, China, and the EU has been the object of debate in the food versus fuel dilemma, both crop area and average yields in the Russian Federation, for example, still show considerable gaps as compared to their former levels.

It is furthermore surprising that the concept of increasing harvesting frequency (increasing the average number of harvests per unit of arable land) did not receive more attention in the discussion on land-use claims by biofuel production. Publications by FAO have reported openly on this phenomenon, which also has been observed in the past.

Finally, it is recommended to make use of biomass balances while assessing issues such as undernutrition and the way it may be affected by biofuel production. The large amount of food that is lost in comparison to the amount of food crops used in biofuel production poses a serious analytical issue. Also, much more attention should be given to the amount of food that would be required to solve hunger rather than just counting the number of undernourished people.

250   *J.W.A. Langeveld, J. Dixon and H. van Keulen*

References

Charles, D. (2009). Corn-based ethanol flunks key test. *Science*, Vol 324, pp587–588.

Cotula, L., Dyer, N., and Vermeulen, S. (2008). *Fuelling exclusion? The biofuels boom and poor people's access to land*. London: International Institute for Environment and Development (IIED); Rome, Italy: FAO.

EIA. (2013). Energy Information Agency. Retrieved 1 February 2013 from http://www.eia.gov/cfapps/ipdbproject

FAO (2011). Feeding the world. Retrieved 13 November 2013 from: http://www.fao.org/docrep/015/i2490e/i2490e03a.pdf

FAO. (2013). FAOSTAT. Retrieved 1 February 2013 from http://faostat.fao.org

Golub, A.A., and Hertel, Th. W. (2012). Modeling land-use change impacts of biofuels in the GTAP-BIO framework. *Climate Change Economics,* Vol 3, p30.

Hunger Notes. (2013). World hunger poverty facts and statistics: Does the world produce enough food to feed everyone? Retrieved 1 February 2013 from http://www.worldhunger.org/articles/Learn/world%20hunger%20facts%202002.htm#Does_the_world_produce_enough_food_to_feed_everyone

Lawson, S., and MacFaul, L. (2010). *Illegal logging and related trade: Indicators of the global response*. London: Chatham House.

Macedo, M.N., DeFries, R.S., Morton, D.C., Stickler, C.M., Galford, G.L., and Shimabukuro, Y.E. (2012). Decoupling of deforestation and soy production in the southern Amazon during the late 2000s. *Proceedings of the National Academy of Sciences,* Vol 109, pp1341–1346.

Stage, J., Stage, J., and McGranahan, G. (2009). *Is urbanization contributing to higher food prices?* London: IIED; New York, USA: UNFPA.

Wegner, L., and Zwart, G. (2011). *Who will feed the world? The production challenge*. Oxfam Research Report. Oxford, UK: Oxfam.

16 Outlook

J.W.A. Langeveld, H. van Keulen
and J. Dixon

16.1 Introduction

In Chapter 15 we saw how enabling policies triggered biofuel production increases, raising production by 70 billion litres of ethanol and 14 billion litres of biodiesel between 2000 and 2010. The speed at which this has happened is remarkable, and targeted biofuel expansion largely has been realised. There are, however, large differences in profitability of biofuel plants. These differences are mostly caused by changes in feedstock prices, which have shown high variations partly because of the linkages between food and energy markets. At a given moment, on average one in three or four plants may be either inactive or running below optimal capacity. This applies mostly to corn ethanol plants in the USA and biodiesel plants around the world.

In particular, the ethanol industry in the USA and biodiesel chains in the EU have been developing very fast. This is not only explained by effective policies but also by favourable market conditions. More prudent policies have been implemented in China, the Far East, and southern Africa, while in Brazil, after decades of strong support for the industry, recent conditions for the expansion of ethanol production have been rather unfavourable (e.g., government tax, high sugar prices, and relatively poor weather conditions).

Policy objectives suggest that growth will remain in place for the coming years. Major players such as Brazil, the USA, and the EU have formulated objective targets that require continuous strong growth. It remains to be seen if these optimistic goals can be realised. Much will depend not on availability of land or crop feedstocks but, rather, on economic conditions, especially for ethanol and biodiesel plants.

Another potential disrupting factor could be weather, which can affect crop production; for example, in recent years droughts were observed in Australia, the Russian Federation, and, more recently, North America. Provided that weather and market changes remain within certain limits, a continuation of strong biofuel production growth can be expected. Previously we saw how changes in land-use intensity in major biofuel producers have fostered substantial increases in crop production within a relatively short time

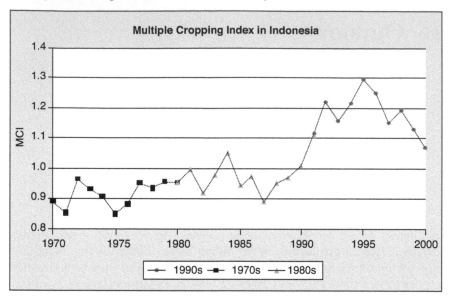

Figure 16.1 Multiple Cropping Index in Indonesia (1970–2000)
Source: Calculated from FAOSTAT (2010–2013), http://faostat.fao.org.

period. While these developments may appear to be exceptional, they are not limited to biofuel producers and are not uncommon.

For example, consider the changes in Multiple Cropping Index (MCI) estimated for Indonesia in the period 1970–2000 (Figure 16.1). Following the introduction of an ambitious irrigation investment programme linked to subsidies for farm inputs (seeds, fertilisers, and agro-chemicals) – the so-called INMAS program (Intensifikasi Massal or mass intensification) – MCI increased in the period from the first oil crisis until the Asian financial crisis from 0.85 in 1975 to 1.30 in 1995.

Consequently, in 20 years crop intensity has increased by more than 50%. For a country such as Indonesia, with nearly 25 million ha of arable land, this is equivalent to an expansion of harvested area of 12 million ha (almost similar to the net biofuel area increase in the entire study area over a decade). After 1995, MCI has declined again.

A decline of MCI has also been observed after the collapse of the Soviet Union. Agricultural area, arable area, and harvested area intensity in the Russian Federation strongly declined after the break up of the Soviet Union (Figure 16.2). More than 6 million ha of agricultural land has been lost since 1992, while a further 4 million ha of arable area was transformed into permanent grasslands. During a period of six years, MCI in Russia lost one-third of its value. In the 1990s its value fell from 0.87 to 0.62. Since reaching

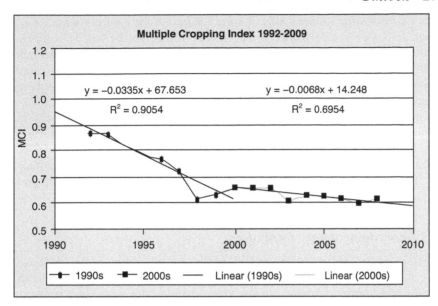

Figure 16.2 Changes in Multiple Cropping Index in the Russian Federation in the period 1992–2009

Source: Calculated from FAOSTAT (2010–2013), http://faostat.fao.org.

this value in 1998, small increases have dropped again after 2004. Because the Russian Federation has 120 million ha of arable land, a difference of 0.10 MCI value represents an area of harvested crops equal to 12 million ha – again, nearly the net increase in biofuel area in the period 2000 to 2010. If we include changes in other former Soviet republics (e.g., Ukraine, central Asian republics), the effect is even more significant.

Major policy and market changes in the past have not only caused shocks in land-use intensity but also provoked strong changes in crop yields. This was observed when the EU (which at the time included 15 member states) introduced market regulation aimed to reduce food surpluses. Farmers responded by improving crop yields (cultivating less land with higher yields was more profitable), and surplus land was abandoned. It was also imminent when farmers in former east bloc countries such as Poland dropped fertiliser use almost overnight after the fall of the Berlin Wall.

These changes, some of which may have been largely overlooked by researchers studying potentials of land-use dynamics related to biofuel policies, clearly define how crop production systems may respond to policy and market changes. They provide a background against which the changes in land use and crop production that have been observed in this book can be evaluated.

In the following sections this chapter evaluates the environmental, economic, and social performance of biofuel production in the study countries. Results from individual chains presented in earlier chapters are brought together, and outcomes are compared in a systematic way. The results are used to draw generic conclusions. Special attention is given to the major points of concern for biofuel production that were presented in the first chapter.

The analysis starts with an evaluation of environmental impacts of biofuel production (Section 16.2), followed by an economic assessment (Section 16.3), evaluation of the impacts for food production (Section 16.4), and a social assessment (Section 16.5). After that, we discuss policy (Section 16.6) before presenting a discussion (Section 17.6) and some chapter conclusions (Section 16.8).

16.2 Environmental performance

As presented in Table 16.1, crop yields are high for sugarcane (Brazil, South Africa), sugar beet, and oil palm. Cereal yields are high for corn in the USA but less so for corn and wheat in the EU and China. Rapeseed and soybean yields are modest. Ethanol yields are highest for sugar beet and sugarcane (Brazil). The highest biodiesel yields were observed for oil palm. Co-product yields are high for corn (USA), oil palm, and sugar beet. Yields are low for rapeseed and soybean, while no co-products for feed markets are generated by sugarcane ethanol.

Table 16.1 Crop, biofuel, and co-product yields of crop production chains

Region	Feedstock	Crop Yield (tonne/ha)	Biofuel Yield (litre/ha)	Biofuel Yield (GJ/ha)	Co-product Yield (tonne/ha)
Brazil	Sugarcane	79.5	7,200	152	–
Brazil	Soybean	2.8	600	18	1.8
USA	Corn	9.9	3,800	80	4.2
USA	Soybean	2.8	600	18	1.8
EU	Wheat	5.1	1,700	37	2.7
EU	Rapeseed	3.1	1,300	43	1.7
EU	Sugar beet	79.1	7,900	168	4.0
Indonesia/ Malaysia	Palm oil	18.4	4,200	90	4.2
China	Corn	5.5	2,200	46	2.9
China	Wheat	4.7	1,700	36	2.5
Mozambique	Sugarcane	13.1	1,100	23	–
South Africa	Sugarcane	60.0	5,000	107	–

Sources: Crop yields calculated from FAOSTAT (2010–2013), http://faostat.fao.org; biofuel and co-product yields taken from Chapters 6–12.

Changes of input productivity and impact on soil organic matter (SOM) are presented in Table 16.2. Crop productivity (expressed in kg of dry matter per kg of nitrogen fertiliser) is low in the EU and China for wheat and rapeseed, but high in Brazil; it is also high for soybean in the USA and for beets in the EU. Biofuel production per kg of nitrogen is low in the EU, China, and Indonesia and Malaysia. The highest production is realised by soybean. Sugarcane and sugar beet take an intermediate position. Biofuel production per kg of phosphate is high for sugarcane, corn, oil palm, and sugar beet, but low for rapeseed and wheat in Europe. Water productivity varies strongly. The lowest water requirements are reported for sugar crops and the highest are reported for oil palm, soybean and – to a lesser extent – rapeseed.

Many chains have a negative impact on SOM – notable exceptions are sugarcane and soybean in Brazil. Corn and oil palm may have a positive SOM balance provided mineralization is limited. SOM may be maintained on soils with low SOM stocks: clay soils for corn and non-peat soils for oil palm. Feedstock management factors such as no-till, burning of sugarcane

Table 16.2 Efficiency and impact of biofuel crop production systems

Region	Feedstock	Input Productivity (crop level)		Input Productivity (fuel level)		Use of Agro-chemicals	Impact on SOM
		kg dm/ kg N	GJ/ kg N	GJ/kg P_2O_5	m^3 of water/GJ	kg a.i./ha	–
Brazil	Sugarcane	327	2.2	1.9	37	0.4	Positive
Brazil	Soybean	595	13.6	1.4	145	12.4	Positive
USA	Corn	50	0.9	1.7	73	2.4	Ambiguous
USA	Soybean	238	5.4	1.4	145	1.6	Negative
EU	Wheat	18	0.5	1.0		3.5	Negative
EU	Rapeseed	16	0.4	0.8	100	3.0	Negative
EU	Sugar beet	165	2.3	1.8	23	4.0	Negative
Indonesia/ Malaysia	Palm oil	61	0.5	1.7	200	4.9	Ambiguous
China	Corn	20	0.4	0.8		–	Negative
China	Wheat	18	0.3	0.6		–	Negative
Mozam-bique	Sugarcane	196	1.3	1.7	50	–	Positive
South Africa	Sugarcane	168	1.1	1.4	50	–	Positive

a.i. = active ingredient

Source: Chapters 6–12.

leaves, and removal of crop residues will strongly affect the impact of the chain on SOM. Current developments, such as mechanical harvesting of sugarcane in Brazil (which no longer requires burning of leaves before harvest) and increased no-till practices (conservation agriculture, as applied in Brazil and the USA), lead to enhanced availability of crop residues and reduced decomposition rates and increase the soil organic matter and thus carbon balance.

Removal of sugar beet leaves and of corn stover will reduce the amount of organic matter that is left in the fields. The risks of soil carbon depletion can be limited by harvesting only part of the available residues (currently being propagated in the USA's corn belt) and by application of sequential cropping: growing another crop ('catch crop') after harvest of the main crop to catch available nutrients and incorporate them into organic material. Catch crops are compulsory in some cases (e.g., when growing maize in soils sensitive to leaching in the Netherlands). Conservation agriculture is one of the few proven sustainable agriculture approaches that has been successfully demonstrated in many different regions and farming systems. While there is complementarity with first-generation ethanol and biodiesel production, there could be competition for biomass when second-generation lignocellulosic fuel systems become common.

Input efficiencies presented in Table 16.2 are well within the ranges for C3 and C4 crops given in Chapter 3. Many biofuel feedstocks are among the best responding crops with respect to input use in agriculture. They are chosen because of their efficient past performance and their ability to deliver large amounts of cheap biofuel feedstocks. Cane, corn, oil palm, and – although used to a lesser extent – sugar beet are more productive crops than many of their alternatives.

Soil quality

Input-use efficiency is largely determined by crop physiology and management. As a rule, higher efficiencies can be realised in better soils. Figure 16.3 shows the impact of soil type on fertiliser-use efficiency. For corn in the USA and wheat in the EU, we calculated how much fertiliser nitrogen could be replaced if farmers could make better use of nitrogen released during mineralisation. Because crop nitrogen requirements do not fully synchronise with mineralisation – the latter is mostly driven by temperature and moisture and tends to continue after harvest – maximum uptake is set at 50%. Figure 16.3a suggests that high-clay soils in the USA can replace up to 250 kg of nitrogenous fertiliser, thus more than current application levels. Low-clay soils can generate up to 150 kg of nitrogen. Differences for EU wheat cultivation, comparing a typical clay soil with a sandy soil as often is found in the eastern part of the continent, are even larger.

The impact on fertiliser efficiency is evident. Utilising nitrogen released by mineralisation allows for a reduction in fertiliser application levels, which

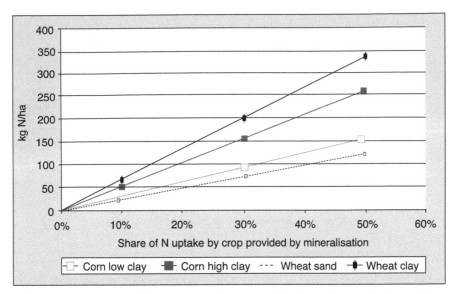

Figure 16.3a Impact of clay content and N mineralisation on N uptake: (high) clay soils require lower fertiliser application levels.

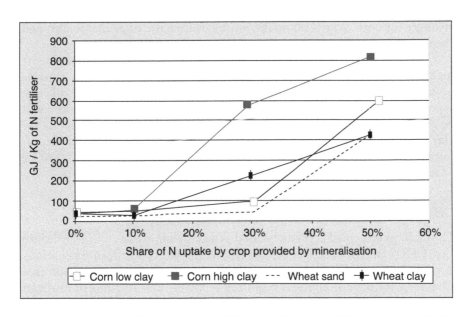

Figure 16.3b Impact of clay content and N mineralisation on N productivity: (high) clay soils show higher crop N productivity at similar uptake shares.

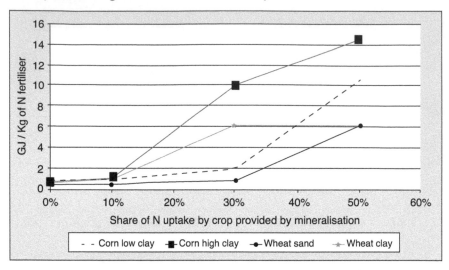

Figure 16.3c Impact of clay content and N mineralisation on N productivity: (high) clay soils show higher biofuel N productivity at similar uptake shares.

will lead to higher dry-matter production (Figure 16.3b) and biofuel energy produced (Figure 16.3c) per unit of nitrogen applied. Reactions are stronger for soils with more clay particles, which tend to offer more protection to soil organic matter and, therefore, build up larger soil carbon stocks. They are stronger for corn, which can make more effective use of available nitrogen (being a C4 crop).

Differences between high- and low-clay soil types tend to decline at higher shares of released nitrogen uptake as high-clay soils reach a maximum amount of fertiliser replaced. Similar effects are expected for all crop types, be it that the exact response will be crop specific. Differences will be more extreme for oil palm, which is either cultivated on peats or on clay soils. Under normal conditions (i.e., draining soils during cultivation), release of nutrients from decaying peat material can reach up to 2,000 kg of nitrogen.

GHG emissions

The findings presented here can be used as a guideline in assessing greenhouse gas (GHG) emissions of biofuel chains. As the use of nitrogen fertilisers is a major source of GHG emissions during crop and biofuel production (see Chapter 4 for details), biofuels requiring relatively low amounts of nitrogen fertilisers (thus realising high GJ yields per kg of nitrogen) will generally show more favourable GHG balances. This is less apparent for soybeans (which require little nitrogen fertiliser but demonstrate very low fuel production levels per ha); it also partly explains the effective GHG balances of sugarcane and sugar beet.

As was discussed earlier, fertile soils that are high in organic matter – which can provide considerable amounts of nutrients – require less fertiliser. Consequently, the use of relatively more productive soils is an effective way to enhance GHG performance of any biofuel chain. A word of warning, however, is needed against replacing artificial fertilisers with animal manure. Manure is a rich source of nitrogen and other nutrients, but it is associated with high to very high levels of N_2O emissions, a strong GHG 310 times more harmful than CO_2.

One consequence of the large increases in MCI reported in the previous chapter is related to the amount of land-use change that would be caused. Because net area for food crops has increased substantially there is no reason to assume that biofuel expansion directly caused (or increased) competition for scarce land resources – although many observers have noted some effect on commodity prices. It is concluded that these findings argue *against* the concept of indirect land-use change (ILUC) following the production of biofuel feedstocks; increased use of food crops in biofuel production does not by definition cause ILUC.

This is surprising and will certainly lead to discussions, especially as most studies so far seem to agree that biofuel expansion causes ILUC. In the EU and USA emissions caused by ILUC already have been formally integrated into biofuel sustainability legislation. As it was phrased during a workshop held at the European Parliament on 20 February 2013: ILUC is a reality. All models and studies now show that there is an impact and that this is above zero. The debate about ILUC is, however, ongoing and new studies are offering new insights especially as to how large the impact would be. Interestingly, more recent research reports seem to generate lower ILUC estimates as compared to earlier work (compare, for example, outcomes presented by Darlington et al. [2013] to those reported by Laborde et al. [2011] and, especially, Searchinger et al. [2008]).

Summarising, it is not probable that biofuel production automatically causes ILUC. It may provoke shortening of fallow or increased integration of grasslands in crop rotations, and that may lead to enhanced carbon releases. But the assumption that future enhanced biofuel production in, for example, Brazil will automatically cause the conversion of nature (forest, savannah) or permanent grassland in this region can no longer be supported.

There are two important reasons for delinking biofuel feedstock production and ILUC. First, in most regions increases in MCI have been generating additional crop harvested areas that exceed net biofuel expansion (which means that there are other reasons – not related to biofuels – that have triggered conversion of nature area to cropping). Second, there is sufficient available agricultural land (e.g., in former Soviet republics or in countries with large grassland areas) that can readily support the expansion of crop cultivation.

Land dynamics are extremely complex, and biofuel expansion is not the only cause of declining crop areas (or even the largest reason). Data presented earlier show that increasing cropping intensity has been more than compensating for the loss of food area due to biofuel expansion. The question then is what is making farmers convert forest and grasslands into agricultural land? Is it really the perception of growing food demand driven by a (perceived) shortage of agricultural land (we have seen that, in fact, harvested area has increased and not declined)? Or are there more logical reasons for conversion of forestland (such as the expected profit that can be made from the sale of timber or from grazing livestock on freshly opened – hence 'free' – forest soils)?

Earlier chapters have shown that the amount of harvested area has increased not only in many individual countries but also in the study area in general, so the rationale for expansion of agricultural land in nature areas should be reviewed again. Farmers may respond to price changes or expected price improvements, but it is unclear why this is not occurring elsewhere, for example, in former Soviet bloc countries.

If GHG emissions related to ILUC are adjusted downward, in principle all biofuel GHG balances would be improved. Depending on the way ILUC emissions originally were calculated, the impact can be high to very high. Excluding these emissions will make future emission reduction objectives (50% or 60%) more feasible for biodiesel producers.

16.3 Economic aspects

Many biofuel policies are linked to economic support measures for the agricultural sector, which is necessary in so far as biofuels are often more costly than their fossil counterparts. It is difficult to assess how much subsidy is involved. The highest estimate we found refers to the USA, where US$7 billion was spent in 2010, and similar estimates were reported for the EU (IISD/GSI, 2013). Figures for other countries were hard to obtain, but given the amounts that apparently have been spent in Brazil (US$11 billion over the period 1970–2000), this is expected to be much smaller.

On the surface, the level of support might appear to be quite high, and it is understandable that some analysts have objected to the support received by biofuels. However, it should be recognized that production agriculture subsidies, which are orders of magnitude larger, and fossil fuel industries also benefit from public subsidies. Subsidies for fossil fuels amounted to US$412 billion in 2010 and have increased to US$523 billion per year in 2011. In this light, biofuel subsidies seem to be modest. Further, subsidies for agricultural production seem also to be in the range of US$300–400 billion per year. In discussing economic support to biofuel production, support in fossil energy and agriculture should also be considered.

16.4 Competition for food

In the previous chapter it was demonstrated that increased use of biomass for biofuel production was greatly exceeded by enhanced agricultural crop production (notably double cropping). Consequently, the availability of arable crops for food and feed markets has increased. It can be concluded that during the time span that is covered (the first decade of the twenty-first century), expansion of land used for domestic biofuel production did not directly significantly affect food production.

Nearly half of biofuel feedstock energy is recovered in co-products. Soybean meal, palm meal, and rapeseed cake make excellent animal feeds, which have been used for many decades. Dried distillers grains with solubles (DDGS) is relatively new as feed. Its digestibility is limited to cattle, but it is the subject of much research and early results are promising. Biomass use for biofuels in the study area amounted to about 527 million tonnes in 2010, up from 193 million in 2000. Some 93 million tonnes of co-products are recovered. Consequently, net biomass use in biofuels increased by 254 million tonnes.

Production intensity is increasing (Figure 16.4). Production per unit of arable land increased by 1.5% in the 1990s but has showed a spectacular annual 13% increase since 2000. This has contributed considerably to biomass availability. It shows, once again, that there has been no real direct need to expand significantly agricultural area on behalf of increased biofuel

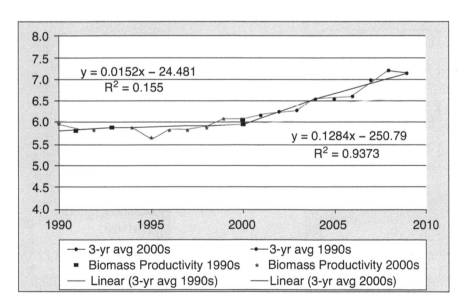

Figure 16.4 Two decades of Biomass Productivity

feedstock demand. Intensification has been larger than the estimated area of biofuel expansion in the studied countries. Increased MCI corresponded to 42 million ha in the study area and more than 90 million ha worldwide.

In terms of biomass, production in the study area grew by more than 870 million tonnes between 2000 and 2010. This has been more than sufficient to compensate for the net biomass use in biofuel production, leaving room for an increased biomass use in food production of more than 600 million tonnes in 2010.

Not all of this is attributed to increasing cropping intensity or yield improvement. As discussed in earlier chapters, there is a third explanation for the huge improvement of crop biomass output: replacement of less productive crops by higher-yielding alternatives. Following the enhanced demand for biomass (be it for food, feed, fibre, or fuels), farmers tend to switch to more productive alternatives. In some cases, crops in the rotation can be replaced. The impacts of this can be huge. Consider, for example, replacing 1 ha of barley in the USA by a similar unit of corn, which yields 6 tonnes more corn than barley plus 6 tonnes more stover (Figure 16.5). Complete biofuel and co-product chain generates more biomass for food and feed markets than barley (not including biofuel production; 4.2 tonnes of DDGS, compared to 3.9 tonnes of barley).

Crop replacement can be considered as a way farmers respond to the increased demand for biomass. It has helped to reduce tensions on food and feed (non-biofuel) crop markets. The net impact of biomass

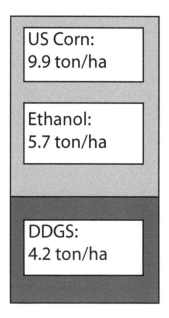
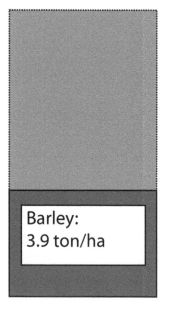

Figure 16.5 Biomass production and use of barley and ethanol-corn in the USA

productivity (BP) improvement suggests that non-biofuel crop availability did not suffer. This does not mean that biofuel production has not affected food biomass availability. Note that the current analysis is restricted to domestic production of biomass feedstocks. Two factors need to be considered here. First, major biofuel-producing regions, such as the USA and the EU, could rely mostly on imported biomass. Second, in theory, they could reduce food exports as an instrument to make sure that sufficient biomass is available for the biofuel industry. A comparison of crop import–export balances in Brazil, the USA, and the EU, however, suggests that no major changes have occurred in balances of major cereals, sugar crops, or oil crops. Reported increases in non-biofuel availability are mostly due to increased domestic crop production.

16.5 Social implications

The popular perceptions of biofuels sometimes are not positive, and are often related to the idea that biofuel production is one of the main factors pushing up food crop prices. However, the fact that they help to reduce prices for transportation is often ignored.

Biofuel production industries in countries such as Brazil, Germany, and the USA can offer valuable options for employment; for example, providing temporal employment for workers who have few other alternatives in Brazil (e.g., Sawyer, 2008). However, manual harvesting – one of the most demanding jobs in rural Brazil – is being phased out by mechanized alternatives. In Germany the number of jobs in the renewable industry has been estimated at 380,000. The industry is also known for its innovative character. Research in the USA suggests that income generated by farmers selling biofuel feedstocks is subject to considerable multiplier effects, which benefit rural non-farm economies (Gan et al., in press).

Land rights are crucial in determining implications of biofuel production. Because most biofuel feedstock production systems are based on large-scale land-use operations, risks for local inhabitants are huge, especially where land rights are not formalised on paper deeds, and so on. For this reason, some authors have emphasised the need to consider not only the biophysical potential of biofuel production, but also the socioeconomic conditions – especially patterns of land ownership (Rosegrant et al., 2008).

16.6 Discussion

During the first decade of the new century, biofuel production has shown a substantial expansion, especially in Brazil (ethanol), the USA, the EU, and China. A continuation of biofuel growth can be expected notably in Brazil and the USA. Recent policy changes are expected to slow down domestic production in the EU, while in China and Indonesia and Malaysia production growth has already weakened. Elsewhere, more efforts will be needed to realise existing policy objectives (e.g., for Brazilian biodiesel production and in southern Africa).

There is much confusion about the amount of land that has been allocated to biofuel production so far and the amount that remains to be used in the near future. There are several reasons for this. First, although the amount of land in use in 2010 locally is considerable (13 million ha in the USA and 9 million ha in the EU), our analysis suggests that net increases in biofuel area (14 million ha) between 2000 and 2010 are much lower. In many studies, co-products are not (sufficiently) covered in the analysis, thus leading to overestimations of biofuel areas. In reality, while the increase locally has been considerable – especially in the USA and, to a lesser extent, in the EU, Brazil, and China – it has been very modest elsewhere. Biofuel area is also expected to have grown in Argentina.

A second reason for confusion is the fact that most analysts refer to partial land balances only. Changes in urban areas and expansion of forest, touristic, infrastructural, or other land use, therefore, tend to be ignored. This is easily leading to an overemphasis of the role of biofuel expansion in land-use change.

Third, changes in cropping intensity seem to be largely overlooked. While projections by FAO indicate that most (90%) production increases are expected to come from a combination of yield improvement and increasing cropping intensity (Nachtergaele et al., 2012), models used for the evaluation of land-use change seem to include yield effects only.

The combination of these three omissions has had serious implications through the way the impact of biofuels has been projected to affect land-use changes. This is very unfortunate, as it has influenced the reputation of biofuels (and that of other renewables), which has generally been viewed in a negative light. It has resulted in calls to change biofuel policies, challenging the ability of biofuels to add to sustainably generate energy – even suggesting that they were worse than the fossil fuels they are to replace.

It is surprising to see to what extent the introduction of biofuel policies have been projected to affect land-use patterns. Changes in agricultural production and land use between scenarios depicting biofuel policies and business-as-usual scenarios tend to be relatively small (e.g., 3.5% extra production of biofuel crops due to the implementation of EU biofuel policies, as reported by Banse et al., 2010). In a recent International Food Policy Research Institute study, Al-Riffai et al. (2012) calculated a change in arable crop area of 0.07% (again caused by EU biofuel policies), which is equivalent to only 1 million ha.

Changes that are projected seem remarkably small. Consider, for example, the impact of EU biofuel production on land use in Brazil. By the year of 2020, 0.5 million ha would be used to support biofuel exports to the EU. Thus, during a period of 20 years, only 0.8% of Brazil's arable area (58 million ha in 2000) would be allocated. Only 15% of that would cause a conversion of primary forest, thus amounting to 0.072 million ha. This represents 0.3% of forest area that was lost in Brazil in the period 2000–2010.

The total ILUC area projected seems to be no more than 2 million ha, but this relatively small amount of land-use change would cause so much carbon release that biofuel GHG emission reduction levels would be seriously affected.

We discussed earlier, however, that biofuel expansion is much smaller than the amount of additional crop area that is generated by increasing cropping intensity, which makes it unlikely that biofuels are a major cause of land-use change.

16.7 Conclusion

Biofuel production showed major expansion during the first decade of the twenty-first century, and the growth is by and large expected to remain during the next few years. In the EU corn and wheat feedstocks are expected to account for more than 80% of the expansion of corn ethanol. Ethanol production in Brazil is also expected to increase substantially. The USDA projects that sugarcane ethanol production in this country will grow considerably up to the year 2020 (USDA, 2013), primarily to meet increasing domestic demand for transportation fuel with higher ethanol blends. However, exports to the EU and USA are also expected to increase. Corn-based ethanol production is expected to double in Argentina by 2022. In Canada ethanol production is expected to increase by 35%, with corn imports accounting for an increasing share of the feedstock.

According to the USDA, in 2012 China used 4.6 million tonnes of corn and 1 million tonnes of wheat to produce ethanol. Due to policies to limit the expansion of grain- and oilseed-based biofuel production, no significant expansion is expected.

Regarding US corn production, the USDA projects that corn acreage will remain high in the near term, with normal yields leading to an increase in production and recovery of corn use. According to the USDA, after several years of adjusting markets, increasing producer returns are expected to lead to gradually increasing corn acreage after 2015. While projected increases in corn-based ethanol production are expected to be much smaller throughout the next decade, ethanol will remain a strong presence in the sector. The report notes that approximately 35% of total corn use is expected to go to ethanol production through 2022.

At the global level, remarkable resilience has been demonstrated. While increasing population combined with economic prosperity – mostly, but not only, in Asia – global crop production has shown that it is able to enhance crop output on often shrinking agricultural resources (USA, Europe – especially former eastern bloc nations). Response has been overwhelming. Increased multiple cropping globally generated 90 million ha in just 10 years. That is immense and exceeds by more than four times the expanded land claims by biofuels.

Meanwhile, it is important to monitor the sustainability of biofuel feedstock production. In this book we have given special attention to efficiency of input use and impacts of crop production on soil and water. We introduced several performance indicators. Crop production systems involved in biofuel production appear to be among the most productive and often also the most efficient. Cane in Brazil (low fertiliser inputs, high yields, efficient nutrient recycling, positive impacts on soil carbon, no irrigation) is one of the best examples. But also other major feedstocks show encouraging performance. US corn yields are among the highest for cereals per ha, while input use is generally fair. EU wheat rape seems to suffer from luxury high input-use levels. EU beets are more efficient. Oil palm in Indonesia and Malaysia is one of the most effective systems, provided it utilises mineral soils.

The analysis presented in this book generally was based on more or less standard data on production conditions. However, variation in soil characteristics is high. It is stressed here that the way in which soil conditions are integrated in crop management is crucial not only for determining yield levels and farm incomes but also for sustainability aspects of production. Our analysis suggests that current practices leave ample room for improvement.

This book presented and analysed data on land resources, biofuel policies, biofuel production, and feedstock cultivation and conversion. It also assessed efficiency of input use and discussed implications for biomass availability, land-use change, and economic performance, as well as social implications. The outcomes sometimes have been surprising. Mostly, it linked available (but sometimes not very accessible) data and knowledge. This has allowed us to make analyses which led to some non-obvious conclusions and new insights. Some key contents and conclusions can be drawn.

- The analysis of biofuel production chains was based on a more extensive description of prevailing local conditions referring to land resources, climate, crop type and crop production systems, and biomass-to-biofuel conversion practices. This information has helped to gain a better insight into the performance and perspectives of biofuel production. As a result, this book includes an improved link of GHG calculations to cropping systems, conversion technology, and soil type.
- Existing assessments seem to under-value co-products that are generated during biofuel production. As a result, the influence of biofuel production on biomass availability for food and feed applications tends to be over-estimated.
- New approaches were presented to assess the impact of biofuel production on biomass availability and land use. This includes the distinction between land cover and land use. Changes in area harvested should not be literally translated into changes in land cover. Loss of agricultural land due to urbanisation generally has been neglected in discussions and analysis.

- Reported insights provide food for thought with respect to assessing how farmers respond to major changes in policy and market conditions. Apparently, responsibility has been much stronger than was generally anticipated.
- It is likely that farmers have responded to especially higher commodity prices. This seems to have been the major incentive behind the increased multiple cropping (i.e., MCI) and biomass output per ha of arable land (i.e., BP). This has had a strong impact on the way biofuel expansion affected land-use change (i.e., ILUC).
- With respect to productivity and input efficiency, it appears that the conditions under which crops are cultivated are much more important than previously taken into account in debate and analysis. This is most clearly demonstrated by differences in the performance of sugarcane in Brazil as compared to other regions (e.g., southern Africa) or wheat in the EU as compared to China.
- Impacts of land preparation seem to receive very limited attention in sustainability assessments. This leads to an underestimation of the way farmers can adapt to (new) requirements for (sustainable) production (consider the impact of no-till on soybeans in Brazil and the USA, and low-till and no-till on corn in the USA).
- The influence of soil type in biofuel feedstock production is often neglected while the impact can be huge. For example, compare the performance of peat versus mineral soils in the oil palm industry in the Far East, or the performance of high vs low clay soils in Figure 16.3.
- Comparison of biofuel support to agricultural and fossil energy support suggests that the former is not exceptionally high. Having said that, for Brazil and the USA support seems to be assisting industry development very well.
- Not all effects of biofuel expansion can be measured directly. Indirect effects include reduction of poverty in Brazil due to large economic growth, reduction of costs for the military in fossil-producing regions of the USA and the EU, and so on.

16.8 Recommendations

This book has presented a large amount of data, some of which leads to different conclusions than those that are often drawn. It is essential that more information is taken into account; for example, when evaluating the impact of biofuel production on land use and changes therein.

The following recommendations can be made:

- The impact of soil conditions and processes on the performance of crop production are huge but remain underestimated. It is recommended that soil properties are included in crop cultivation practices and assessments of sustainability performance. This holds especially for nutrient application and productivity assessments, soil carbon management, and carbon balances and soil water management and efficiency evaluation.

- Co-products generated by biofuel production chains should receive more attention in the analysis of biofuel policies. Failing to acknowledge the full dimension of co-product generation will lead to an underestimation of the outcomes of biofuel production and an overestimation of land impacts.

- New concepts need to be developed and implemented with respect to the dynamics of crop production, land cover, and land use (harvested area). Changes in harvesting intensity – reflected in MCI – should be incorporated in the analysis of land-use change (i.e., ILUC).

- Land impacts should be based on the analysis of complete land balances and should be differentiated by type of broad farming system. Generally, loss of agricultural land due to expansion of non-agricultural (non-biofuel) economic activities should be better integrated in the biofuel impact assessments.

- Changes in cropping intensity are critical underpinning for rational policy development. Thus, international agricultural statistics urgently need to estimate and publish estimates on MCI. (This should be reflected in the economic modelling work supporting impact assessments of biofuel policies.)

- The performance and the prospects of biofuel production require a profound understanding of the different farming systems contexts; that is, the patterns of crop and livestock production and livelihoods and the interactions with feedstock production. Thus, it is recommended that biofuel Life Cycle Assessment (LCA) analyses consider not only the feedstock field but also the whole farm context.

- Economic models that are used in the evaluation of biofuel policies should be making more use of updated and relevant biophysical information in their descriptions of crop production practices.

- Evaluation of biofuel policies should be done against the background of agricultural and (fossil) energy policies that are currently in place. Thus, governments need to assess the complementarity and coherence of policies in related sectors with a view to harmonization of policies. Any assessment of biofuel subsidies, for example, should also consider existing fuel subsidies and their impact on energy prices and the sustainability of energy production.

- Recommendations for (changes in) biofuel policies seem to put little emphasis on the importance of a stable policy environment. More continuity in policy will reduce production and price uncertainty and encourage biofuel producers to optimize biofuel production.

- More attention is needed on the position of smallholders, land owners, indigenous people, and other rural groups whose lives and way of living sometimes are threatened by expansion of large-scale crop production. This is equally relevant to expanding non-biofuel production (e.g., of soybean for animal feed in Latin America or oil palm for human food and cosmetics consumption in the Far East).

• Sustainability assessments should – more than they currently do – be based on information of local biofuel feedstock production conditions. Crop rotations, prevailing soil types, land management, and input use and other systems aspects all have a major impact on environmental performance of biofuel chains and the way farmers can make use of management options to realise given economic and sustainability objectives in different farming systems.

References

Al-Riffai, P., Dimaranan, B., and Laborde, D. (2012). *Global trade and environmental impact study of the EU biofuels mandate*. Washington, DC: International Food Policy Research Institute.

Banse, M. van Meijl, H. and Woltjer, G. (2010). *Biofuel Policies, production, trade and land Use*. In: H. Langeveld et al. (eds.), The biobased economy. Biofuels, materials and chemicals in the post-oil era (pp244–258). London, UK: Earthscan.

Darlington, Th., Kahlbaum, D., O'Connor, D., and Mueller, S. (2013). *Land use change greenhouse gas emissions of European biofuels policies utilizing the Global Trade Analysis Project (GTAP) model*. Macatawa., USA: Air Improvement Resource.

Gan, J., Langeveld, J.W.A., and Smith, T. (in press). Biomass producer decision making: Direct and indirect transfers in different spheres of interaction. Accepted for publication in *Environmental Management*.

IISD/GSI. (2013). *Addendum to biofuels: At what cost? A review of costs and benefits of EU biofuel policies*. Geneva, Switzerland: The International Institute for Sustainable Development; Global Subsidies Initiative.

Laborde, D. (2011). *Assessing the land use change consequences of European biofuel policies: Final report*. Washington, DC: International Food Policy Research Institute.

Nachtergaele, F., Bruinsma, J., Valbo-Jorgensen, J., and Bartley, D. (2011). *Anticipated trends in the use of global land and water resources*. Rome, Italy: Food and Agricultural Organization of the United Nations.

Rosegrant, M.W., Ewing, M., Msangi, M., and Zhu, T. (2008). *Bioenergy and global food situation until 2020/2050*. Berlin, Germany: Wissenschaftlicher Beirat der Bundesregierung Globale Umweltveränderungen. Retrieved from http://www.wbgu.de/wbgu_jg2008_ex08.pdf

Sawyer, D. (2008). Climate change, biofuels and eco-social impacts in the Brazilian Amazon and Cerrado. *Philosophical Transactions B*, Vol 363, pp1747–1752.

Searchinger, T., Heimlich, R., Houghton, R. A., Dong, F., Fabiosa, J., Tokgoz, S., Hayes, D., and Hsiang, Yu, T. (2008). Use of U.S. croplands for biofuels increases greenhouse gases through emissions from land-use change. *Science*, Vol 319, pp1238–1240.

USDA. (2013). *Agricultural projections to 2022*. Retrieved 19 February 2013 from www.usda.gov/oce/commodity/projections/

Index

.